Some Abstract Algebra

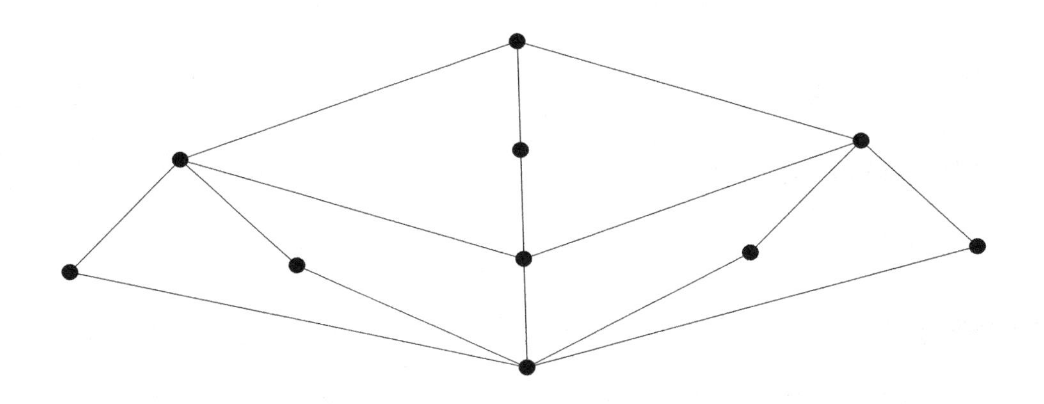

A Primer and Interactive Workbook

Richard Grassl - Tabitha Mingus

Richard Grassl
Emeritus Professor of Mathematical Sciences
University of Northern Colorado
Greeley, Co 80639
richard.grassl@unco.edu

Tabitha Mingus
Associate Professor, Collegiate Mathematics Education
Mathematics Department
Western Michigan University
tabitha.mingus@umich.edu

Summer 2019 Edition

A current electronic version can be found for free at
http://www.openmathbooks.org/someabstract/

Prepared for publication by Oscar Levin for Open Math Books

Preface

This book consists of two parts: one, a primer designed to provide an adequate introduction to the essentials of abstract algebra and to some related number theory, and two, a workbook designed to enable the reader to interactively engage with colleagues in exploring the fascinating world of abstract algebra.

We have taken a problem solving approach – the primer alone contains over 130 problems. So be prepared for minimal text material to read, combined with worksheets that extend and enhance text topics. These worksheets are designed to encourage discovery of interesting relationships between algebraic structures, geometry, mappings, and proofs.

Very little, if any, background in abstract algebra is needed for a course based on this Primer and the workbook. This material has been used successfully for over a decade with in-service secondary teachers seeking licensure or an MA degree in teaching mathematics.

In this book we embrace the oft-quoted maxim - "You learn mathematics by doing mathematics." Such an effort leads to better understanding and deeper learning.

Finally, a valuable by-product: A significant number of teachers who have studied this material have incorporated a variety of the worksheets into their secondary curriculum as they encounter topics like closure, binary operations and their properties, modular arithmetic, and the structure of the integers (yes, GCD and LCM show up), and the rational and real numbers.

Richard Grassl

April 2019

Some Abstract Algebra - A Primer

and some number theory

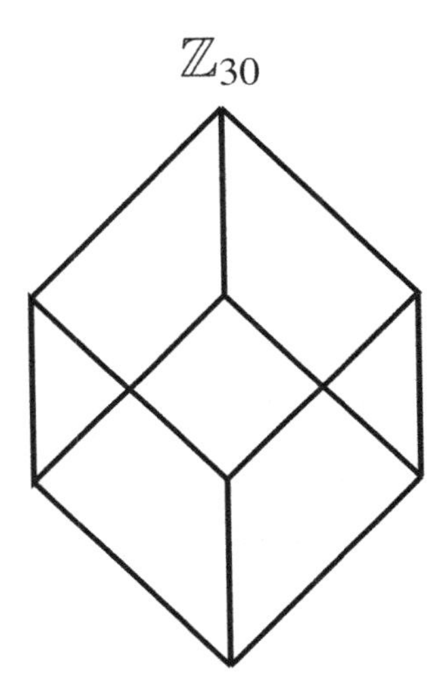

Richard Grassl
University of Northern Colorado
Emeritus Professor of Mathematical Sciences
richard.grassl@unco.edu

Edited and typeset in LaTeX by Conner Hatton

Contents

Introduction

An abstract algebra consists of a set of objects (integers, real numbers, permutations, polynomials, matrices, ...), various binary operations, along with some properties (closure, inverses, commutativity,...). Examples of abstract algebras include groups, rings, integral domains, and fields. Operations include rotations of regular geometrical figures, ordinary and modular addition and multiplication, addition and multiplication of matrices and of polynomials, composition of permutation cycles, direct products and others.

In one sense the core ideas of algebra are abstracted out and viewed from a much larger lens. For example, the problem of finding analogues of the quadratic formula, around the mid 1500's, led to the study of the symmetric groups which shed light on the nonsolvability of the general quintic.

Applications are plentiful. Among the many fields of study making significant use of algebraic structures we include cryptography, genetics, mineralogy, the study of molecular structures in chemistry, elementary particle theory in physics, Latin squares in statistical experiments, and finally, architecture and art.

Important contributors over the past several centuries include Joseph Lagrange, Niels Abel, Arthur Cayley, Emmy Noether, Gauss, Galois, Sylow among many others.

Binary Operations

The concepts of being commutative and associative are usually introduced to students as they study the four basic arithmetic operations of addition, subtraction, multiplication and division of integers. These four operations denoted by $+, -\times, \div$ are examples of binary operations.

Let S be any set. A binary operation on S is a function $f : S \times S \longrightarrow S$. Sometimes a binary operation is depicted using the infix notation $(m,n) \mapsto m \,\Box\, n$, rather than the prefix notation $f(m,n)$. The following are examples of binary operations on $\mathbb{N} = \{0, 1, 2, 3, \ldots\}$.

OPERATION	INFIX NOTATION
$f(m,n) = m + n$	$m \,\Box\, n = m + n$
$f(m,n) = mn$	$m \,\Box\, n = mn$
$f(m,n) = \gcd(m,n)$	$m \,\Box\, n = \gcd(m,n)$
$f(m,n) = 5^m \cdot n$	$m \,\Box\, n = 5^m \cdot n$

Additional binary operations that will be of importance include

$$f(x,y) = x \div y \text{ on } S = \mathbb{R} \text{ where } y \neq 0$$

$$f(m,n) = m - n \text{ on } S = \mathbb{Z} = \{0, \pm 1, \pm 2, \ldots\}$$

$$f(m,n) = m + n - mn \text{ on } S = \mathbb{Z}$$

$$f(m,n) = m + n - 1 \text{ on } S = \mathbb{Z}$$

$$f(A,B) = A \cup B \text{ on } P(S) \text{ for } S = \{\text{a,b,c}\}$$

$$f(A,B) = A \cap B \text{ on } P(S) \text{ for } S = \{\text{a,b,c}\}$$

Here, $\cup$ denotes union and $\cap$ denotes intersection. Also, $P(S)$, the power set for S, is the set of all subsets of S. As an example, if S={a,b},then $P(S)$ = { $\emptyset$, {a}, {b}, {a,b} }.

PROPERTIES OF BINARY OPERATIONS

Binary operations on a set X may or may not satisfy the following properties:

Commutativity: $x \square y = y \square x$ for all x, y in X.

Associativity: $x \square (y \square z) = (x \square y) \square z$ for all x, y, z in X.

Identity: An element $e \in X$ such that $x \square e = e \square x = x$ for x in X is called an identity for the binary operation $\square$.

Inverses: If e is an identity under $\square$, an inverse of an element a in X is an element b in X such that $a \square b = e = b \square a$.

For example, the operation + on $\mathbb{Z}$ is associative since $a + (b + c) = (a + b) + c$ for all a, b, c in $\mathbb{Z}$. Since $a + b = b + a$, + is commutative. The element 0 serves as an identity e since $0 + a = a + 0 = a$. Each element $a \in \mathbb{Z}$ has an inverse, namely $-a$.

One important characterization or consequence of the notation $f : A \times A \longrightarrow A$ is that the result $f(m, n)$, or $m \square n$, must be an element in A; i.e., A must be closed under the binary operation $\square$. So divisibility, denoted $\div$, is not a binary operation on $N = \{0, 1, 2, \ldots\}$ since $m \div n = \frac{m}{n}$ is not necessarily an integer. But $\div$ is a binary operation on the set $\mathbb{R}^+$ of positive real numbers. Notice that since $2 \div 3 \neq 3 \div 2$, $\div$ is not commutative.

Example 1

Consider the binary operation $m \square n = 3^m \cdot n$ on $\mathbb{N} = \{0, 1, 2, \ldots\}$; show that $\square$ is not associative. The following single example accomplishes this:

$$1 \square (0 \square 1) = 1 \square 1 = 3 \text{ but } (1 \square 0) \square 1 = 0 \square 1 = 1$$

Does $\square$ have an identity? it might be natural to try 0 or 1. Since $0 \square n = n$ but $n \square 0 = 0$, 0 is not an identity. Since $1 \square n = 3n$, 1 is not an identity. No other element in n works either; without an identity, there are no inverses.

Example 2

Union, $\cup$, is a binary operation on $P(S)$ where S={a,b,c}. Since $A \cup (B \cup C) = (A \cup B) \cup C$, $\cup$ is associative. Since $\emptyset \cup A = A = A \cup \emptyset$ for all $A \in P(S)$, the empty set $\emptyset$ serves as an identity.

When the set A is finite, a binary operation $\square$ can be given by a matrix table where the element $x \square y$ is found at the intersection of row x and column y.

Example 3

Let $A = \{1,3,7,9\}$ and let $x \square y$ be the digit in the units position upon ordinary multiplication of x and y. This is sometimes written as $x \square y \equiv xy(\text{mod } 10)$. The matrix table is

$\square$	1	3	7	9
1	1	3	7	9
3	3	9	1	7
7	7	1	9	3
9	9	7	3	1

This binary operation $\square : A \times A \longrightarrow A$ is associative (you need to check $4 \cdot 4 \cdot 4 = 64$ cases), is commutative (from the symmetry of the table), has identity 1, and each element has an inverse as shown in the following table:

x	1	3	7	9
inverse of x	1	7	3	9

Example 4

The binary operation $\circ$ on A={e, a, b} given by the table below is associative, commutative and has e as an identity. In the exercises you are asked to find inverses.

$\circ$	e	a	b
e	e	a	b
a	a	b	e
b	b	e	a

Example 5

Let $A = \{1, 2, 5, 10\}$ and $m \square n = \gcd(m, n)$, the greatest common divisor of m and n. The matrix table for $\square$ is given. With some effort, you can show that $\square$ is associative.The table's symmetry verifies that the operation $\square$ is commutative. In the exercises, you are asked to determine an identity and to see if there are inverses.

$\square$	1	2	5	10
1	1	1	1	1
2	1	2	1	2
5	1	1	5	5
10	1	2	5	10

Example 6

Ordinary multiplication with rounding to the nearest tenth after each multiplication is not associative. Try this one: (1.1)(0.3)(2.7).

Example 7

Define a binary operation $\circ$ on $\mathbb{Z}$ by $m \circ n = m + n - 3$. To find the identity e, e needs to satisfy $m \circ e = e \circ m = m$. Since $\circ$ is clearly commutative we need not worry about checking $e \circ m = m$. Then, $m \circ e = m + e - 3 = m$ and $e = 3$. As an example, $7 \circ 3 = 7 + 3 - 3 = 7$. What about inverses? The inverse of an integer p in $\mathbb{Z}$ must satisfy $3 = p \circ p^{-1} = p + p^{-1} - 3$ and so $p^{-1} = 6 - p$. As an example, the inverse of 13 is -7 since $13 \circ (-7) = 13 + (-7) - 3 = 3$, the identity.

"An" Identity versus "The" Identity

Throughout this discussion of properties we have been saying "an" identity. Here is some good news! We can replace "an" with "the" whenever an identity exists.

Theorem 1

If the binary operation $\square$ on X has an identity, then it is unique.

Proof: Proceed using a proof by contradiction. Suppose there are two different identities; call them e and f. Since $x \square e = x = e \square x$ for all x in X, it must hold for $x = f$, i.e., $f \square e = f = e \square f$. Since $x \square f = x = f \square x$ for all x in X, it must hold for $x = e$, i.e., $e \square f = e = f \square e$. But then $e \square f = f$ and $e \square f = e$, or $e = f$ contradicting the fact that they were different.

Similarly, inverses are unique. Let e be the identity for an associative operation $\circ$ on X, and let g and h be two inverses for some a in X. Then $g = g \circ e = g \circ (a \circ h) = (g \circ a) \circ h = e \circ h = h$. You should give reasons for each step.

Meet and Join

There are two binary operations on $B=\{0,1\}$ that are basic in computer design and operation. The <u>meet</u> $\wedge$ and <u>join</u> $\vee$operations are given by the tables below.

$\wedge$	0	1
0	0	0
1	0	1

$\vee$	0	1
0	0	1
1	1	1

The meet operation is similar to intersection $\cap$ and join is similar to union $\cup$, and behave like the logical connectives "and" and "or" respectively. Each of $\bigwedge$ and $\bigvee$ are commutative and associative. Are there identities, inverses?

Bitwise Addition Modulo 2

Another binary operation having applications in coding theory is based on the table below.

$\oplus$	0	1
0	0	1
1	1	0

Here, $a \oplus b$ is 0 if $a + b$ is even and $a \oplus b = 1$ if $a + b$ is odd. Equivalently, $a \oplus b = 0$ if $a = b$, and $a \oplus b = 1$ if $a \neq b$. The binary operation $\oplus$ on $B = \{0, 1\}$ is called "bitwise addition modulo 2". On $B^2 = \{00, 01, 10, 11\}$, the table for $\oplus$ is given.

The operation is performed bitwise; $10 \oplus 01 = 11$ since $1 \oplus 0 = 1$ and $0 \oplus 1 = 1$. You are asked to investigate properties of $\oplus$ in the exercises.

$\oplus$	00	01	10	11
00	00	01	10	11
01	01	00	11	10
10	10	11	00	01
11	11	10	01	00

PROBLEMS - BINARY OPERATIONS

1. (a) Is subtraction a commutative binary operation on $\mathbb{Z}$? Explain.

   (b) Is subtraction associative on $\mathbb{Z}$?

   (c) Is multiplication commutative on $\mathbb{Z}$?

   (d) Does multiplication have an identity on $\mathbb{Z}$? Are there inverses?

2. Give an example of subsets A and B of $\mathbb{Z}$ so that

(a) subtraction is not a binary operation on A.

(b) multiplication is not a binary operation on B.

3. Let S={a,b,c}, and $P(S)$ be the power set of S.

   (a) Is $\cup$ on $P(S)$ commutative Explain.

   (b) Does {a} have an inverse?

4. Let S={a,b,c,d}.

   (a) Is $\cap$ on $P(S)$ commutative, associative? Explain.

   (b) Does $\cap$ have an identity?

   (c) What is the inverse of A={b,d} under $\cap$?

5. Let $\circ$ be the binary operation on $\mathbb{N} = \{0, 1, 2, \ldots\}$ with $m \circ n = (5^m)(2n+1)$.

   (a) Compute $2 \circ 3$ and $3 \circ 2$. Is $\circ$ commutative?

   (b) Is $\circ$ associative? Does $\circ$ have an identity? Is it 2-sided?

6. Let $\circ$ be the binary operation $m \circ n = m + n - mn$ on $\mathbb{Z}$. Is $\circ$ commutative, associative? Does $\circ$ have an identity?

7. Make a table of inverses for the operation in example 4.

8. Let $A = \{1, -1, i, -i\}$ and let $\square$ denote ordinary complex multiplication.

   (a) Make the matrix table for $\square$. (b) Is $\square$ associative, commutative?

   (c) Does $\square$ have an identity?

   (d) Give a table of inverses for the elements of A.

9. What is the identity for the binary operation in example 5? Are there inverses?

10. Let $A = \{1, 2, 5, 10\}$ and define a binary operation on A by $m \square n =$ LCM(m, n). Make the matrix table for $\square$ and decide whether $\square$ is commutative, has an identity, inverses.

11. Let $A = \mathbb{Z} = \{0, \pm 1, \pm 2, \pm 3, \dots\}$ and define a binary operation on $\mathbb{Z}$ by $m \square n = m + n - 1$.

    (a) Is $\square$ commutative? (b) Does $\square$ have an identity?

    (c) Are there inverses?

12. Define the operation $\square$ on $\mathbb{Z}^+$ as follows: $m \square n = \gcd(m, n) + \text{lcm}(m, n)$. Is $\square$ associative?

13. (a) Is $m \square n = 3^m \cdot n$ commutative on $\mathbb{N} = \{0, 1, 2, 3, \dots\}$?

    (b) How many ordered pairs (m, n) can you find so that $m \square n = 18$?

14. Which properties are satisfied by $\oplus$, bitwise addition modulo 2?

15. Show that $m \square n = n\, 2^m$ is not associative on $\mathbb{N} = \{0, 1, 2, \dots\}$. Is $\square$ commutative?

16. Let S be the set of all real numbers except -1 and define $a \square b = a + b + ab$ on S. What is the identity? What is the inverse of an element p in S?

17. Let $a \square b = a^b$ with $a, b \in \{1, 2, 3, \dots\}$.

    (a) Is $\square$ commutative? (b) Is $\square$ associative?

    (c) What is the meaning of a^{b^c}?

Activity 1 - AN OPERATION TABLE

Here is the operation table for a binary operation □. Is □ associative? Is □ commutative?

□	*a*	*b*	*c*	*d*	*e*
a	*a*	*b*	*c*	*d*	*e*
b	*b*	*c*	*a*	*e*	*c*
c	*c*	*a*	*b*	*b*	*a*
d	*b*	*e*	*b*	*e*	*d*
e	*d*	*b*	*a*	*d*	*e*

CLOSURE

We say that a binary operation $\square$ on a set S is CLOSED if whenever a and b are any two elements in S then $a \square b$ is also in S. For example, the operation $+$ is closed on the set $S = \mathbb{Z} = \{0, \pm 1, \pm 2, \ldots\}$ since $a + b$ is in $\mathbb{Z}$ for $a, b \in \mathbb{Z}$. But subtraction is not closed on the set $\mathbb{N} = \{0, 1, 2, 3, \ldots\}$ since $1 - 3 = -2$ is not in $\mathbb{N}$.

The following gives the two part procedure for determining if a particular set S is closed under a particular binary operation $\square$:

1. Choose two arbitrary elements for S; label them a, b.

2. Show that $a \square b$ is a number of S.

Example 1

The set $E = \{0, 2, 4, 6, \ldots\}$ is closed under addition. Let $a = 2m$ and $b = 2n$ be arbitary elements of E. Then since $2m + 2n = 2(m + n)$ is an even integer the sum of $2m$ and $2n$ is in E. E is also closed under multiplication since $(2m)(2n) = 4mn = 2(2mn)$ is even.

Example 2

The set of all rational numbers of the form $3^n, n \in \mathbb{Z}$, is closed under multiplication since $3^a \cdot 3^b = 3^{a+b}$.

PROBLEMS - CLOSURE.

In each of the following if the set is closed under the operation give reasons (actually a proof); if not, provide a counterexample.

1. Is $A = \{0, 1, 4, 9, 16, \ldots\}$ closed

a. under addition? b. under subtraction? c. under multiplication?

2. Is $B = \{0, \pm5, \pm10, \pm15, \ldots\}$ closed

   a. under addition? b. under multiplication?

3. Is $C = \{0, 2, 4, 6\}$, a finite set, closed under addition?

4. Is $D = \{1, 3, 5, 7, \ldots\}$ closed

   a. under addition? b. under multiplication?

5. Is $E = \{1, 4, 7, 10, 13, \ldots\}$, the positive integers having remainder 1 upon division by 3 closed

   a. under addition? b. under multiplication?

6. Repeat #5 with $F = \{2, 5, 8, 11, 14, \ldots\}$.

7. The set $G = \{0, \pm4, \pm8, \pm12, \ldots\}$ is closed under subtraction. Give another set H that is closed under subtraction. Show that $G \cap H$ is also closed under subtraction.

8. Is the set of all rational numbers of the form 2^n, where n is in $\mathbb{Z}$, closed under multiplication?

9. Is the set of all positive rational numbers closed under addition? Under multiplication?

10. Let $I = \{2^m \cdot 3^n : m, n \text{ are in } \mathbb{Z}\}$. Is I closed under multiplication?

    Hint: Is $3/8$ in I? Is $1/9$ in I?

11. Are the irrationals closed under multiplication?

12. Prove that if S and T are sets of integers closed under subtraction so is the intersection $S \cap T$. Is the union of S and T also closed under subtraction?

13. Why does 0 always have to be a member of any set that is closed under subtraction?

14. Let R denote a 120° rotation of an equilateral triangle. Is the set $\{I, R, R^2\}$ closed under "rotation"? Here, I means do nothing and R^2 means a 240° rotation.

15. (a) Express each of 5 and 13 as a sum of two squares. Express 65 as a sum of two squares.

    (b) Is the set $A = \{m^2 + n^2 : m, n \in \mathbb{Z}\}$ closed under ordinary multiplication?

Units Multiplication

Under units multiplication the product of any two positive integers, denoted by $a \square b$, is the units digit of the product under ordinary multiplication. So $5 \square 9 = 5$, $7 \square 8 = 6$.

16. Is the set $\{0, 2, 4, 6, 8\}$ closed under units multiplication?

17. Is the set $\{1, 3, 7, 9\}$ closed under units multiplication?

18. Is $\{1, 4, 6\}$ closed under units multiplication? How about the set $\{2, 4, 8\}$? How about $\{1, 5, 9\}$?

19. What would you have to add to the set $\{1, 3, 5\}$ to make it closed under units multiplication?

Closing Comments: As you can see from the problems, the concept of closure is important. Analyzing for closure promotes a deeper understanding of the use of counterexamples. It also facilitates moving from specific examples to general results.

Activity 2 - THE PENNY MOVE

Suppose you allow a penny to move in just the following four ways on this square.

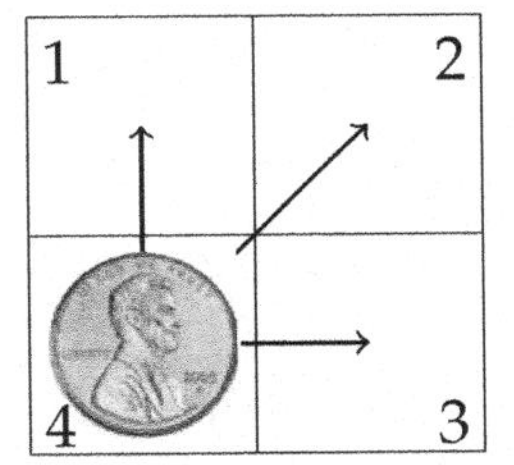

1. I the penny stays stationary.
2. H the penny can move horizontally left <u>or</u> right.
3. V the penny can move vertically, up <u>or</u> down.
4. D the penny can move diagonally.

<u>TASK 1</u> - Starting in box 3, where does the penny land after H? After V?

<u>TASK 2</u> - Starting in box 1, where does the penny land after D is followed by V?

<u>TASK 3</u> - Where is the penny? Start in box 2 and then do the following (left to right) $DDVHIDH$.

TASK 4 - Repeat with the sequence $HDDHDVIHV$, but start in box 3.

TASK 5 - Fill in the operation table giving the result of a move followed by another move. The binary operation $\circ$ is "followed by".

(a) Does it matter where you start?

(b) Is V followed by H the same as H followed by V?

$\circ$	I	H	D	V
I				
H				D
D				
V				

The D that is shown indicates that an H followed by a V gives a D. Always operate from the left-most column to the top row.

(c) Is the "Penny move" operation closed?

(d) What observations can you make about your chart?

Groups

A GROUP is an algebraic structure that consists of two items: a set of elements G, and a binary operation $\circ$. This structure satisfies FOUR axioms:

CLOSURE: For any elements a and b in G, $a \circ b$ is also in G.

IDENTITY: There is a unique element e in G such that for any $a \in G$ we have $a \circ e = a = e \circ a$.

INVERSES: For every element a in G there is an element a^{-1} in G so that $a \circ a^{-1} = e = a^{-1} \circ a$

ASSOCIATIVITY: For any three elements a, b, c in G we have $(a \circ b) \circ c = a \circ (b \circ c)$.

A general group can be denoted as $(G, \circ)$ indicating the importance of having both a carrier set G and a binary operation $\circ$. When the operation is clear from a particular context we may write just G for that group.

Example 1

$(\mathbb{Z}, +)$ is a group under the usual addition operation. Choose a, b in $\mathbb{Z}$. Since $a + b$ is an integer CLOSURE holds. The IDENTITY e is 0 since $0 + a = a = a + 0$ for any $a \in G$. The INVERSE of a is $-a$ (we could write $a^{-1} = -a$) since $a + (-a) = 0 = (-a) + a$. ASSOCIATIVITY holds since $(a + b) + c = a + (b + c)$.

Example 2

In the exercises you will show that the following are groups: $(\mathbb{Q}, +)$, $(\mathbb{R}, +)$, $(\mathbb{Q}^+, \times)$, $(\mathbb{R}^+, \times)$ where $\mathbb{Q}$=Rationals, $\mathbb{R}$=Reals, $\mathbb{Q}^+$= Positive Rationals, $\mathbb{R}^+$=Positive Reals.

Example 3

In the exercises you will show that the following are not groups: $(\mathbb{Z}, -)$, $(\mathbb{Z}, \div)$. What do you think goes wrong with the binary operation $\div$?

Abelian Groups

Some groups have an additional fifth property called commutativity. A binary operation $\circ$ on a set G is commutative if $a \circ b = b \circ a$ for all a, b in G. We also say that the group $(G, \circ)$ is an ABELIAN GROUP, named after Niels Abel, a major contributor in the development of group theory. He also proved the insolvability of the fifth-degree polynomial equation, one of his greatest achievements.

Caution: The words commutative and abelian are almost synonymous. In doing proofs, one can never say "G is abelian since it is commutative". Often commutative describes a binary operation while abelian describes a group.

The examples involving $\mathbb{Z}$, $\mathbb{Q}$, and $\mathbb{R}$ above are all abelian groups.

Group Tables

When the set G is finite (most of the above examples were infinite) the four group properties and be readily detected from an operation table. We saw earlier that the set $G = \{1,3,7,9\}$ was closed under units multiplication denoted $\otimes$. The operation table follows.

$\otimes$	1	3	7	9
1	1	3	7	9
3	3	9	1	7
7	7	1	9	3
9	9	7	3	1

The symbol $\otimes$ means take a from the left most column and "multiply" by b from the very top row. The 16 interior elements are just 1, 3, 7 and 9 showing closure. The element 1 acts like the identity – look at the top interior row and the left most interior column. Inverses are easy to find; just look for the 1's in the table. Since $1 \otimes 1 = 1$, $3 \otimes 7 = 1$ and $9 \otimes 9 = 1$ we have the following table of inverses.

x	1	3	7	9
x^{-1}	1	7	3	9

Associativity is inherited from multiplication in $\mathbb{Z}$; we now know that $(G, \otimes)$ is a group. In fact, the symmetry of the table shows that G is an abelian group.

Example 4

Here is an example of a group that involves functions and algebra. The operation is composition of functions. $Let f(x) = x$, $g(x) = \frac{1}{1-x}$ and $h(x) = \frac{x-1}{x}$. $G = \{f, g, h\}$ is a group with identity function f. The "product" gh is $g(h(x)) = g(\frac{x-1}{x}) = \frac{1}{1-(\frac{x-1}{x})} = \frac{x}{x-(x-1)} = x$. Conclusion: $gh = f$. You should form the other products and make the group table.

Consequences of the 4 Group Axioms

Theorem 1

If $ab = ac$ in a group G then $b = c$. (This is called left cancellation).

Proof: Multiply each side of $ab = ac$ on the left by a^{-1}. You get $(a^{-1}a)b = (a^{-1}a)c$ or $b = c$. A similar result holds for right cancellation. But be careful "mixed" cancellation may not work, i.e. $ab = ca$ does not necessarily imply $b = c$.

Theorem 2

In the multiplication table for a finite group $G = \{g_1, g_2, \ldots g_n\}$ each element of G appears exactly once in each row.

Proof: The entries on row r are $rg_1, rg_2, rg_3, \ldots, rg_n$. If two of these are the same, say $rg_i = rg_j$ then $g_i = g_j$ upon left multiplication by r^{-1}. But this is a contradiction. Why? It can also be shown that elements in any column are distinct.

Theorem 3

For any $a \in G$, $(a^{-1})^{-1} = a$.

Proof: $a a^{-1} = e$. This is like saying that the inverse of 2 in $(Q, \times)$ is $2^{-1} = \frac{1}{2}$ since $2 \cdot \frac{1}{2} = 1$, the identity.

Theorem 4: Sock-Shoe Theorem

$(ab)^{-1} = b^{-1}a^{-1}$.

Proof: $(ab)(b^{-1}a^{-1}) = a(bb^{-1})a^{-1} = aea^{-1} = e$.

Theorem 5

If $a = a^{-1}$ for all $a \in G$, then G is abelian.

Proof: $ab = a^{-1}b^{-1} = (ba)^{-1} = ba$. This result is equivalent to saying that if the operation table has e, the identity, down the main diagonal then G is abelian. The reason; $a = a^{-1}$ implies $a^2 = e$.

Theorem 6

Let a and b be in a group G. Show that $(ab)^{-1} = a^{-1}b^{-1}$ if and only if $ab = ba$.

Proof: First we show that if $(ab)^{-1} = a^{-1}b^{-1}$, then $ab = ba$. We have $ab = ((ab)^{-1})^{-1} = (a^{-1}b^{-1})^{-1} = (b^{-1})^{-1}(a^{-1})^{-1} = ba$. Conversely, if $ab = ba$, then $(ab)^{-1} = (ba)^{-1} = a^{-1}b^{-1}$

Theorem 7

Prove that $(ab)^2 = a^2b^2$ in a group G if and only if G is abelian.

The proof of Theorem 7 is left as a problem.

A Group Generator

Sometimes there is an element in a group G whose powers (sums) generate the entire group. In $G = \{1,3,7,9\}$ under units multiplication, 3 is such an element since $3^0 = 1, 3^1 = 3, 3^2 = 9, 3^3 = 7$. We can write $[3] = \{1,3,7,9\}$ to show that 3 is a generator. It is also true that $[7] = [3]$, but $[9] \neq \{1,3,7,9\}$. A group that has a generator is called cyclic. Hence {1,3,7,9} is a cyclic group. $\{1,-1,i,-i\}$ is also a cyclic group under complex multiplication, generated by either i or $-i$. $(\mathbb{Z},+)$ is an additive cyclic group generated by 1 or -1. For an additive group, powers are replaced by sums: $2 = 1+1, 3 = 1+1+1, 4 = 1+1+1+1, \ldots$ and so on. In summary, for a group whose operation is multiplication a^m means $a \cdot a \cdot a \cdot \cdots \cdot a$; if the operation is addition, $m \cdot a$ means $a + a + a + \cdots + a$. Here is a chart showing a comparison between the two notations:

Multiplicative notation	Additive notation
a^{-1}	$-a$
$a = (a^{-1})^{-1}$	$a = -(-a)$
$(ab)^{-1} = b^{-1}a^{-1}$	$-(a+b) = (-a)+(-b)$
$a^m \cdot a^n = a^{m+n}$	$ma + na = (m+n)a$
$a^n = e$	$na = 0$
$[a] = \{a^m : m \in \mathbb{Z}\}$	$[a] = \{m \cdot a : m \in \mathbb{Z}\}$

Subgroups

Let $(G,\circ)$ be a group and let H be a subset of the set G. $(H,\circ)$ is a subgroup of $(G,\circ)$ if $(H,\circ)$ is closed under $\circ$, has the e of G as the identity and contains inverses. Associativity is inherited from $(G,\circ)$. Examples are easy to find.

$(\mathbb{Z},+)$ is a subgroup of $(Q,+)$. {1,9} is a subgroup of {1,3,7,9} under mod 10 multiplication. The set $\{1,-1\}$ is a subgroup of $\{1,-1,i,-i\}$ which itself is a subgroup of all complex numbers under multiplication. The set of all integral multiples of 3, $H = \{0,\pm 3,\pm 6,\dots\}$, is a subgroup of $(\mathbb{Z},+)$. Can you give a subset S of $(\mathbb{Z},+)$ such that S is closed under addition but is not a subgroup?

MODULAR GROUPS

Clock arithmetic provides a fruitful source of nice finite groups. Recalling that 13 o'clock is really just 1 o'clock upon subtraction of 12, we can make a clock with just the four numbers 0,1,2,3 the remainders when any integer n is divided by 4. We can write $7 \equiv 3 (\text{mod } 4)$ for example. This is read as 7 is congruent to 3 modulo 4. In general $a \equiv b(\text{mod } m)$ means that a and b have the same remainder when divided by m; or that $a-b$ is divisible by m.

Definition. Let $\mathbb{Z}_n = \{0,1,2,3,\dots,n-1\}$. The sum $a+b$ of any two elements in $\mathbb{Z}_n$ is just the remainder when $a+b$ is divided by n.

With this definition of sum we can see that $(\mathbb{Z}_n,+)$ is a group. Lets form the operation table for $\mathbb{Z}_4 = \{0,1,2,3\}$.

+	0	1	2	3
0	0	1	2	3
1	1	2	3	0
2	2	3	0	1
3	3	0	1	2

The table shows closure, and that 0 is the identity. Locate the 0's in the table to see that the inverse of 0 is 0, the inverse of 1 is 3 (since 1+3=0), the inverse of 2 is 2, and the inverse of 3 is 1. This new sum rule is an associative binary operation on $\mathbb{Z}_4$ since ordinary addition is associative on $\mathbb{Z}$, the usual integers. In similar analysis, $(\mathbb{Z}_n, +)$ is an additive group.

Example 5

$\mathbb{Z}_6 = \{0, 1, 2, 3, 4, 5\}$ is a group with a number of subgroups. $A = \{0\}$, $B = \{0, 3\}$, $C = \{0, 2, 4\}$, and $D = \{0, 1, 2, 3, 4, 5\}$ are all subgroups of $\mathbb{Z}_6$. The following picture, called a lattice of subgroups, shows the relationships between the subgroups.

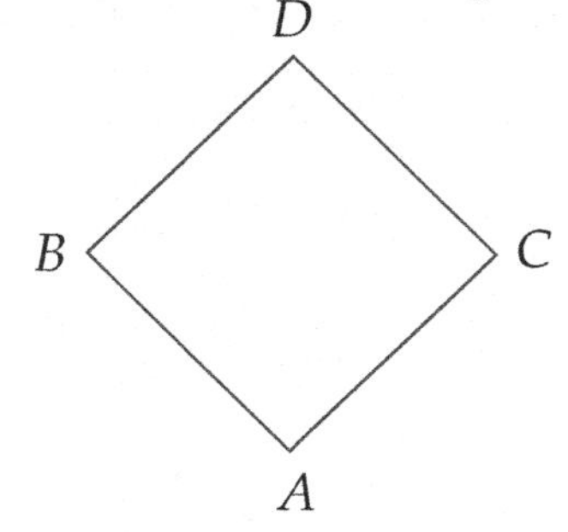

The upward sloping lines indicate subgroup inclusion: $A \subseteq B, A \subseteq C, B \subseteq D, C \subseteq D$ and $A \subseteq B \subseteq D, A \subseteq C \subseteq D$, two chains of length two.

Example 6

The modular groups $\mathbb{Z}_n$ are cyclic where 1 is always a generator for the additive groups $\mathbb{Z}_n$.

For $\mathbb{Z}_4$, 1 is a generator since

$$1 \cdot 1 = 1 = 1$$

$$2 \cdot 1 = 1 + 1 = 2$$

$$3 \cdot 1 = 1 + 1 + 1 = 3$$

$$4 \cdot 1 = 1 + 1 + 1 + 1 = 4$$

In additive notation $3 \cdot 1$ means add three 1's together. In $\mathbb{Z}_6$, only 1 and 5 are generators. For example, 3 is not a generator since you can only make 0 and 3 using multiples $m \cdot 3$ of 3. Try it! Likewise, the element 2 will only generate 0, 2, 4.

The previous example prompts the question: which elements in $\mathbb{Z}_n$ are generators? The following chart might provide a clue as you pursue the question in the exercises.

n	$\mathbb{Z}_n$	generators
2	$\{0, 1\}$	1
3	$\{0, 1, 2\}$	$1, 2$
4	$\{0, 1, 2, 3\}$	$1, 3$
5	$\{0, 1, 2, 3, 4\}$	$1, 2, 3, 4$
6	$\{0, 1, 2, 3, 4, 5\}$	$1, 5$

PROBLEMS - GROUPS

1. Show that each of the following are groups.

   (a) $(Q, +)$ (b) $(R, +)$ (c) $(Q^+, \times)$ (d) $(R^+, \times)$

2. Show that the following are not groups

   (a) $(\mathbb{Z}, -)$ (b) $(\mathbb{Z}, \div)$ (c) $(\mathbb{Z}, \times)$

3. Make the multiplication table for $A = \{4, 8, 12, 16\}$ under multiplication mod 20. Does A have an identity?

4. Is $B = \{2, 4, 6, 8\}$ under units multiplication a group? Is B cyclic? Is B abelian?

5. Verify that 7 is a generator of the group {1,3,7,9} under units multiplication, but that 9 is not a generator.

6. Let $S = \{e, a, b, c\}$. Make the 4×4 group operation table assuming that e is the identity and that $a^2 = b^2 = c^2 = e$. Is S cyclic? Abelian?

7. Show that $G = \{1, -1, i, -i\}$ is a group under ordinary complex multiplication. Is G cyclic?

8. Show that $G = \{00, 01, 10, 11\}$ is a group using bitwise addition mod 2. Is G cyclic?

9. Give an example of a group that illustrates Theorem 6.

10. Prove that in a group $(abc)^{-1} = c^{-1}b^{-1}a^{-1}$. Why is this called the sock-shoe theorem?

11. Make the group table for $G = \{000, 001, 010, 011, 100, 101, 110, 111\}$ using bitwise addition mod 2. Is G abelian? Is G cyclic?

12. Is {1,3} a subgroup of {1,3,7,9} under mod 10 multiplication?

13. Verify that $H = \{0, \pm 7, \pm 14, \ldots\}$ is a subgroup of $\mathbb{Z}$ under addition.

14. Is the set of all complex numbers α with $|\alpha| \leq 1$ a subgroup of all nonzero complex numbers under multiplication? Here, if $\alpha = a+bi$, $|\alpha| = \sqrt{a^2 + b^2}$, its distance from the origin.

15. Make the operation table for $\mathbb{Z}_6$ and find all subgroups.

16. Make the operation table for $\mathbb{Z}_8$ and find all subgroups.

17. Without making the addition table for $\mathbb{Z}_{12}$ can you give all the closed subsets of $\mathbb{Z}_{12}$? Are these in fact subgroups? Draw the lattice of subgroups of $\mathbb{Z}_{12}$.

18. Show that in $(\mathbb{Z}_n, +)$, the additive groups of integers modulo n, that the inverse of any $a \neq 0$ is $n - a$.

19. Explain why $(\mathbb{Z}_4, \bullet)$ is not a group where multiplication is modulo 4.

20. Verify associativity for the following sum in $\mathbb{Z}_7$:

    (3+5)+6 and 3+(5+6)

21. Find all the generators for the cyclic group $\mathbb{Z}_5$ and verify that each in fact generates all of $\mathbb{Z}_5$.

22. Determine those elements in $\mathbb{Z}_n$ that are generators.

23. Solve the quadratic equation $x(x + 1) = 0$

    (a) in $\mathbb{Z}_4$ (b) in $\mathbb{Z}_5$ (c) in $\mathbb{Z}_6$

24. Make the group multiplication table for $G = \{e, a, a^2, a^3, a^4\}$ where e is the identity and $a^5 = e$. Hint: $a^3 \cdot a^4 = a^7 = a^5 \cdot a^2 = a^2$.

25. Let G be a group. Prove that for any $a \in G, H = \{x \in G : x = a^n \text{ for } n \in \mathbb{Z}\}$ is a subgroup of G (generated by a).

26. Prove Theorem 7.

27. Let $\{M_2(\mathbb{R})\}$ be the set of all 2×2 matrices with real entries. Show that this set is a group under matrix addition.

28. Let $G = \{IM_2(\mathbb{R})\}$ consist of all invertible 2×2 matrices. Prove that G is a multiplicative group.

29. Let $G = \{ax + b : a, b \in \mathbb{Z}_2\}$. Prove that G is a group under addition mod 2.

30. Let $G = \{ax^2 + bx + c : a, b, c \in \mathbb{Z}_3\}$. Prove that G is an additive group of order 27.

31. Let a and b be elements of an abelian group, and let n be a positive integer. Prove by mathematical induction that $(ab)^n = a^n b^n$.

32. A certain multiplicative group G, mod 91, has order 9. Which element is missing from the following listing: $1, 9, 16, 29, 53, 74, 79, 81$ of elements of G?

Activity 3 - A CAYLEY TABLE

Complete the following Cayley Group Table

	e	a	b	c	d
e	e	a	b	c	d
a	a	b	c	d	e
b	b				
c	c				
d	d				

1. What is the order of each element?

2. Give the inverse of each element.

Activity 4 - A PARTIAL GROUP TABLE

Complete this operation table.

	e	a	b	c	d	f
e	e	a	b	c	d	f
a	a	b	e	d		
b	b					
c	c	f				a
d	d					
f	f					

More on Cyclic Groups

Recall that if a group G is made up entirely of powers of a particular element, call it a, then G is called a cyclic group denoted by $G = [a]$. The element a is called a generator, and the least positive integer s such that $a^s = e$, the identity in G, is called the order of a. G can be finite or infinite.

Example 1

The additive group Z is cyclic. Either 1 or -1 will generate all of $\mathbb{Z}$. Since $\mathbb{Z}$ is additive, $[-1] = \{m \cdot (-1) : m \in \mathbb{Z}\}$; so $[-1]$ consists of all distinct elements among the "additive powers" $\ldots,$ $-3(-1), -2(-1), -1(-1), 0, -1, -2, -3 \ldots.$

Example 2

$5\mathbb{Z} = [5] = \{0, \pm 5, \pm 10 \ldots\}$ is a cyclic subgroup of $\mathbb{Z}$ generated by 5 (or -5). The generator 5 has infinite order.

Example 3

The order of i in $[i] = \{1, -1, i, -i\}$ is 4, while -1 has order 2.

Example 4

$\mathbb{Z}_n = \{0, 1, 2, \ldots, n-1\}$ is cyclic for all positive n.

$\mathbb{Z}_{12}$ has $1, 5, 7,$ and 11 as generators. Lets check 5:

$0 \cdot 5 = 0$ $\quad 6 \cdot 5 = 6$

$1 \cdot 5 = 5$ $\quad 7 \cdot 5 = 11$

$2 \cdot 5 = 10$ $\quad 8 \cdot 5 = 4$

$3 \cdot 5 = 3$ $\quad 9 \cdot 5 = 9$

$4 \cdot 5 = 8 \qquad 10 \cdot 5 = 2$

$5 \cdot 5 = 1 \qquad 11 \cdot 5 = 7$

You might want to check that 7 also generates $\mathbb{Z}_{12}$. We can write $\mathbb{Z}_{12} = [1] = [5] = [7] = [11]$. Finally, since the gcd$(3, 12) = 3$, the element 3 generates a subgroup of $\frac{12}{3} = 4$ elements; $[3] = \{0, 3, 6, 9\}$

Example 5

Let $G = [a]$ be a cyclic group of order 12. Then with $a^{12} = e$, $[a] = \{e, a, a^2, a^3, \ldots, a^{11}\}$. The generator a has order 12; the element a^7 is also a generator. The order of a^3 is 4 since 4 is the least power s such that $(a^3)^s = e$.

Example 6

Let $G = [a]$ be a cyclic group of order 30. The order of a^9 is 10 and $[a^9] = \{e, a^9, a^{18}, a^{27}, a^6, a^{15}, a^{24}, a^3, a^{12}, a^{21}\}$.The elements are listed in the order that they are made.

The following are two useful results:

Result 1. Let $a \in \mathbb{Z}_n$. $[a] = \mathbb{Z}_n$ if and only if a and n are relatively prime, i.e. $\gcd(a, n) = 1$.

Result 2. Let $G = [a]$ have order n. Then $G = [a^m]$ if and only if $\gcd(m, n) = 1$.

Multiplicative Group of Invertibles

Let V_n denote the subset of the additive group $\mathbb{Z}_n = \{0, 1, 2, \ldots, n-1\}$ of subsets that have multiplicative inverses. Alternatively, we could define V_n as the set of all positive integers less than n that are relatively prime to n. V_n is a group under multiplication modulo n. Showing inverses is the only interesting aspect. If $\gcd(m,n) = 1 = am + bn$, for some integers a and b, then m has an inverse a since $bn \equiv o(n)$, leaving $am \equiv 1(n)$.

Example 7

$V_8 = \{1,3,5,7\}$. The multiplicative identity is 1 and each of $3,5,7$ is its own inverse. V_8 is not cyclic.

Example 8

Each of V_5, V_6, V_7 is cyclic but V_{16} is not. The orders of the elements in $V_{16} = \{1,3,5,7,9,11,13,15\}$ are $1,4,4,2,2,4,4,2$, respectively. Lacking an element of order 8, V_{16} cannot be cyclic.

PROBLEMS - MORE ON CYCLIC GROUPS

1. Is V_9 cyclic?
2. Is V_{15} cyclic ?
3. Let $[a] = \{e, a, a^2, \ldots, a^{23}\}$ be a cyclic group of order 24. List the elements of a subgroup or order 3. What is the order of a^5 in $[a]$?
4. In Example 5, which elements of $[a]$ are generators? Why?
5. Is $G = \{1,3,7,9\}$ under units multiplication cyclic?
6. Prove that every cyclic group is abelian.
7. What is the order of the cyclic subgroup of $\mathbb{Z}_{30}$ generated by 25?

8. What is the order of the cyclic subgroup $[i]$ of the nonzero complex numbers under multiplication?

9. Find the number of generators of a cyclic group with order:

   (a) 7 (b) 9 (c) 15 (d) 60

10. Let $G = [a]$ be a cyclic group of order 18.

    (a) List all the elements of order 3 in G.

    (b) List all the elements of order 4 in G.

11. Let V be the multiplicative group of the nonzero complex numbers, and let $\omega = (-1 + i\sqrt{3})/2$.

    (a) Show that $\omega = \cos 120° + i \sin 120°$.

    (b) Show that $\omega^3 = 1$. What is the order of ω?

    (c) What is the order of $\cos(5\pi/11) + i \sin(5\pi/11)$ in V?

    (d) What is the order of $(1 + i)/\sqrt{2}$ in V?

    (e) What is the order of $1 + i$ in V?

12. Let $G = [a]$ be a cyclic group of order m, generated by a. Can you determine a formula for the order of the element a^r, an element of G?

13. Let $G = \{IM_3(\mathbb{R})\}$, the group of 3×3 invertible matrices with real entries. What is the order of the cyclic subgroup generated by

    (a) $\begin{pmatrix} 0 & 0 & 1 \\ 1 & 0 & 0 \\ 0 & 1 & 0 \end{pmatrix}$ (b) $\begin{pmatrix} 0 & 1 & 0 \\ 1 & 0 & 0 \\ 0 & 0 & 1 \end{pmatrix}$?

WILSON'S THEOREM

The multiplicative group V_n provides us with a structure that yields a proof of an interesting number theoretic result - WILSON'S THEOREM. John Wilson(1741-1793) is usually credited with the theorem, although many feel that some credit belongs to Lagrange.

Theorem 1: Wilson's Theorem

Let $p \neq 2$ be a positive prime. Then $(p-1)! \equiv -1(\text{mod } p)$. Here is an alternate statement: the product of all the nonzero elements in $\mathbb{Z}_p$ is -1. Or: The product of all the elements in $V_p = \{1, 2, 3, \ldots, p-1\}$ is -1.

Proof: In V_p the product $2 \cdot 3 \cdot \ldots \cdot p-2$ is congruent to 1 mod p. V_p has an even number of elements x such that $x \neq x^{-1}$. Now use $p - 1 \equiv -1(p)$.

Example 1

Let $p = 11; V_{11} = \{1, 2, 3, 4, 5, 6, 7, 8, 9, 10\}$. Nearly all of these elements pair up with their inverses: $2 \cdot 6 \equiv 3 \cdot 4 \equiv 5 \cdot 9 \equiv 7 \cdot 8 \equiv 1$, leaving $1 \cdot 10 \equiv -1$.

Example 2

Let $p = 13; V_{13} = \{1, 2, 3, 4, 5, 6, 7, 8, 9, 10, 11, 12\}$. Here, the product $2 \cdot 3 \cdot 4 \cdot \ldots \cdot 11$ of 10 numbers is congruent to 1, due to the pairing of each x with x^{-1}. This leaves $12 \equiv -1(13)$.

PROBLEMS - WILSON'S THEOREM

1. In V_{13}, pair up the elements and illustrate Wilson's Theorem.

2. Pair up the elements in V_{17} and verify $16! \equiv -1(17)$.

3. Show that $(p-2)! \equiv 1(p)$.

4. What is the remainder when 100! is divided by 101?

5. What is the remainder when 99! is divided by 101?

6. How many solutions to $x^2 = 1$ are there in $\mathbb{Z}_p$?

7. Explain why $(29!)^2 \equiv 1(59)$ and $(30!)^2 \equiv -1(61)$.

Activity 5 - WHEN IS V_n CYCLIC?

For which $n \in \{1, 2, 3, \dots, 20\}$ is V_n cyclic? You might try half of these by hand, and the rest by programming. Can you form any conjectures? If you are ambitious, try extending to $n = 30$ or more.

Activity 6 - IS $(\mathbb{Q}^+, \times)$ CYCLIC?

Is the multiplicative group $(\mathbb{Q}^+, \times)$ of positive rationals under multiplication cyclic?

Activity 7 - A FUN GROUP

The rows and columns of the multiplication table of a group $G = \{a, b, c, d, e, f, g\}$ are headed by the elements in this order. The first five entries in the second row are b, d, f, c, a. Complete the multiplication table.

	a	b	c	d	e	f	g
a	a	b	c	d	e	f	g
b	b	d	f	c	a		
c	c						
d	d						
e	e						
f	f						
g	g						

Lagrange's Theorem

Theorem 1

Let G be a finite group. The order of any subgroup H divides the order of G.

This will be seen by partitioning the elements of G into non overlapping sets called cosets, as illustrated in the following.

Definition. Choose an element a in G. A (left) coset of H is the set aH consisting of all products ah with $h \in H$. The set Ha is a right coset.

Let $H = \{1, 14\}$ be a subgroup of the group of invertibles $V_{15} = \{1, 2, 4, 7, 8, 11, 13, 14\}$. In this example, $1H = \{1, 14\}$, $2H = \{2, 13\}$, $4H = \{4, 11\}$, $7H = \{7, 8\}$. The order of $H = 2 =$ order $G/\#$ of cosets. The number of cosets is called the index of H in G.

In general, $H = \{h_1, h_2, \ldots, h_r\}$, $aH = \{ah_1, ah_2, \ldots, ah_r\}$. Cosets have two important properties:

1. The elements of aH are distinct. If $ah_i = ah_j$, left cancellation will produce a contradiction.

2. Distinct left cosets are disjoint – not too hard to prove. Just assume some element is in both aH and bH and see what happens.

Now let order $G = s$ and order $H = r$. Start making all (left) cosets until you exhaust all elements of G.

$$aH = \{ah_1, ah_2, ah_3, \ldots, ah_r\}$$

$$bH = \{bh_1, bh_2, bh_3, \ldots, bh_r\}$$

$$\vdots$$

Here we take a new $b \in G$ (but $b \notin aH$), and continue this process until all elements in G have been accounted for. There can only be a finite number of these cosets, say t. Then $rt = s$. This completes the proof of Lagrange's Theorem, since r divides s.

Here is an example of an additive group where $o(G)$=12, $o(H)$=4 and the index $t = 3$. Let $G = \mathbb{Z}_{12} = \{0, 1, 2, \ldots, 11\}$ and $H = \{0, 3, 6, 9\}$. The three cosets are

$$0 + H = \{0, 3, 6, 9\}$$

$$1 + H = \{1, 4, 7, 10\}$$

$$2 + H = \{2, 5, 8, 11\}$$

Verify properties 1 and 2 above, and that every element $\mathbb{Z}_{12}$ is in precisely one coset.

As a corollary to Lagrange's Theorem, we have that the order of every element in a finite group must divide $o(G)$.

As a final comment, Lagrange's Theorem tells us that there are really only two "different" groups of order 4, for example. If $o(G) = 4$, then either G is cyclic or not. If G is cyclic, $G = [a]$ where a has order 4. If G is not cyclic then all elements (except e) must have order 2 by Lagrange's Theorem. This latter group is called a Klein 4-group (The penny group is a Klein 4-group).

PROBLEMS - LAGRANGE'S THEOREM

1. Find all the (left) cosets of $H = \{0, 6\}$ in $\mathbb{Z}_{12}$.

2. Find all the cosets of $H = \{0, 4, 8\}$ in $\mathbb{Z}_{12}$.

3. Find all cosets of $H = \{1, 8\}$ in V_9. Does V_9 have any other subgroups?

4. Find all cosets of $H = \{1, 4\}$ in V_{15}. Repeat with $H = \{1, 4, 11, 14\}$.

5. Prove that every group having prime order must be cyclic.

6. Show that a cyclic group of order 22 has one element of order 2 and ten elements of order 11.

7. Let G be a group of order 45. What are the possible subgroup orders?

8. Prove that a group G of order 49 must have a subgroup of order 7. Hint: consider two cases:G is either cyclic or not.

9. Let $G = [a]$ be a cyclic group of order 91. Find a subgroup having index 13.

10. Let K be a subgroup of H, and let H be a subgroup of a group G. If the order of K is 6 and the order of G is 144, what are the possibilities for the order of H?

Here are two important consequences of Lagrange's theorem.

Theorem 2: Fermat's Theorem

If a is an integer and p is a prime, then $a^p \equiv a (\text{mod } p)$.

Proof: Take a look at $V_p = \{1, 2, \ldots, p-1\}$, the multiplicative group of integers mod p. The order of V_p is $p-1$. It then follows that $a^{p-1} \equiv 1 (\text{mod } p)$. Now, multiply by a. (The following more general result holds as a consequence of Lagrange's theorem: if n is the order of a group G and $a \in G$, then $a^n = e$).

Fermat's theorem extends to the following more general result.

Definition. The Euler φ-function. Let $m \in \mathbb{Z}^+$, then $\varphi(m)$ denotes the number of integers a in $\{0, 1, 2, \ldots, m-1\}$ such that $\gcd(a, m) = 1$.

As an example $\varphi(10) = 4$ since the integers $1, 3, 7, 9$ are each relatively prime to 10. $\varphi(11) = 10$ since 11 is prime.

Theorem 3: Euler's Theorem

If $\gcd(a, m) = 1$, then $a^{\varphi(m)} \equiv 1 \pmod{m}$.

Proof: The order of V_m is $\varphi(m)$. It then follows from Lagrange's theorem that

$$a^{\varphi(m)} \equiv 1 \pmod{m}$$

As an example, take $m = 10$ and $a = 3$. First, $\gcd(10, 3) = 1$, and $\varphi(10) = 4$. So $3^4 \equiv 1(10)$.

Activity 8 - THE EULER φ- FUNCTION

$\varphi(n)$ is the number of positive integers less than n that are relatively prime to n. Start by tabulating $\varphi(n)$ for $n = 1, 2, 3, \ldots, 16$. Any conjectures?

Conjecture and prove formulas for

$$\varphi(p), \varphi(p^2), \varphi(p^3), \varphi(p^n), \varphi(pq), \varphi(pqr)$$

How many generators does each of these additive groups have?

$$\mathbb{Z}_p, \mathbb{Z}_{25}, \mathbb{Z}_{27}, \mathbb{Z}_{81}, \mathbb{Z}_{21}, \mathbb{Z}_{105}$$

Prove that $m = \sum_{d|m} \varphi(d)$, where m is a positive integer and the sum is taken over all divisors d of m.

Prove that if $n = p_1^{e_1} p_2^{e_2} p_3^{e_3}$ then $\varphi(n) = n(1 - \frac{1}{p_1})(1 - \frac{1}{p_2})(1 - \frac{1}{p_3})$.

Group Isomorphisms

The set of rational numbers is denoted by $Q = \{\frac{a}{b} : b \neq 0, a, b \in \mathbb{Z}\}$. The special subset of Q where $b = 1$ is essentially like $\mathbb{Z} = \{0, \pm 1, \pm 2, \ldots\}$. For all practical purposes, these two sets are the same. This similarity is addressed in the mathematical language that follows:

Definition. An ISOMORPHISM between two groups $(G, \circ)$ and $(G', \Box)$ is a mapping $\theta : G \mapsto G'$ such that

(a) θ is one-to-one

(b) θ is onto

(c) $\theta(a \circ b) = \theta(a) \Box \theta(b)$ (θ preserves the operation).

For property (c) the operation $a \circ b$ takes place in G, while $\theta(a) \Box \theta(b)$ occurs in the image group G'. This same concept is seen in calculus when we write $\lim(f(x) \cdot g(x)) = \lim f(x) \cdot \lim g(x)$ or $(f + g)' = f' + g'$ for derivatives. It is much more notorious when seen as the students' dream: $(x + y)^2 = x^2 + y^2$ or $\log xy = (\log x)(\log y)$.

Here are several examples.

Example 1

Let $\mathbb{Q}^* = \{\frac{a}{1} : a \in \mathbb{Z}\}$. Then $\mathbb{Q}^*$ is isomorphic to $\mathbb{Z}$ under the isomorphism $\theta(\frac{a}{1}) = a$. θ is $1 - 1$, onto and $\theta(\frac{a}{1} + \frac{b}{1}) = \frac{a+b}{1} = a + b = \theta(\frac{a}{1}) + \theta(\frac{b}{1})$.

Example 2

$\mathbb{Q}^{**} = \{\frac{a}{2} : a \in \mathbb{Z}\}$ is isomorphic to $\mathbb{Z}$.

Example 3

The group $A = \{1, -1, i, -i\}$ under complex multiplication is isomorphic to the cyclic group $B = \{e, a, a^2, a^3\}$ under the following mapping:

x	1	-1	i	$-i$
$\theta(x)$	e	a^2	a	a^3

θ is evidently 1-1 and onto. Here, $\theta(ab) = \theta(a)\theta(b)$ can be checked for each pair a, b. For example, $\theta[i(-i)] = \theta[1] = e$ and $\theta(i) \cdot \theta(-i) = a \cdot a^3 = e$.

Example 4

Let $G = (\mathbb{R}, +)$, the real numbers under addition, and $G' = (\mathbb{R}^+, \times)$, the positive reals under multiplication. The mapping $\theta(x) = 2^x$ is an isomorphism from G onto G'. Properties of logarithms show 1-1 and onto. $\theta(a + b) = \theta(a) \cdot \theta(b)$ follows from $\theta(a + b) = 2^{a+b} = 2^a \cdot 2^b = \theta(a)\theta(b)$.

Example 5

Again, let $G = (\mathbb{R}, +)$. The mapping $\theta : G \longrightarrow G$ (itself) given by $\theta(x) = x^2$ is not an isomorphism.

PROBLEMS - ISOMORPHISMS

1. Find an isomorphism from $(\mathbb{Z}, +)$ to $(2\mathbb{Z}, +)$.

2. Regarding Example 3, is the following also an isomorphism? $\theta(1) = e$, $\theta(-1) = a$, $\theta(i) = a^2$, $\theta(-i) = a^3$.

3. Regarding Example 4, prove that $\theta(x) = 2^x$ is 1-1 and onto.

4. In Example 5, prove that $\theta(x) = x^2$ is not an isomorphism.

5. Can you give an isomorphism from $G' = (\mathbb{R}^+, \times)$ onto $G = (\mathbb{R}, +)$?

6. If group A is isomorphic to the group B, and if A is abelian, prove that B is abelian.

7. If $\theta : G \mapsto G'$ is an isomorphism and e is the identity of G and e' is the identity in G', then $\theta(e) = e'$ and $\theta(x^{-1}) = [\theta(x)]^{-1}$.

8. Let θ be a group isomorphism from G to G', show that if $\theta(a) = a'$ then $\theta(a^n) = (a')^n$.

9. Show that if $\theta(a) = a'$ under a group isomorphism then a and a' have the same order.

10. Prove that V_{10} is isomorphic to V_5, ie, that the set of invertibles in $\mathbb{Z}_{10}$ is isomorphic to the invertibles in $\mathbb{Z}_5$.

11. Let $\theta : (\mathbb{R}^+, \times) \mapsto (\mathbb{R}^+, \times)$ be defined by $\theta(x) = \sqrt{x}$. Is θ an isomorphism from $(\mathbb{R}^+, \times)$ to itself?

12. Show that V_8 is not isomorphic to V_{10}. Hint: Make a table of orders.

13. Show that V_8 is isomorphic to V_{12}.

14. Let $G = \{0, \pm 3, \pm 6, \pm 9, \ldots\}$ and $H = \{0, \pm 7, \pm 14, \pm 21, \ldots\}$. Are G and H isomorphic under addition? If yes, does that isomorphism preserve multiplication?

    From certain of these exercises you can see that essential group properties are preserved under group isomorphisms. We say that these properties are invariant under isomorphisms.

HOW TO PROVE THAT TWO GROUPS ARE ISOMORPHIC:

(a) First produce a mapping θ.

(b) Check that it is 1-1 and onto.

(c) Verify that $\theta(a \circ b) = \theta(a)\square\theta(b)$.

HOW TO PROVE THAT TWO GROUPS G AND G' ARE
NOT ISOMORPHIC:

(a) Show that G and G' do not have the same order.

(b) Show that one is abelian, the other not.

(c) Look at the order of elements of each and note discrepancy.

(d) Show that one is cyclic, the other not.

(e) In general, look at invariants and see if something is not consistent.

Be careful on what you determine to be structural, ie, like the properties just seen in (a) - (e). For example, you cannot say that $\mathbb{Z}$ and $5\mathbb{Z}$ under addition are not isomorphic because 13 is in $\mathbb{Z}$ but not in $5\mathbb{Z}$. That is not a structural property.

15. Prove that $(\mathbb{R}^*, \times)$, the multiplicative group of nonzero real numbers is NOT isomorphic to $(\mathbb{R}^+, \times)$, the positive reals under multiplication. Hint: look for an element of order 2 in each.

16. Prove that $(\mathbb{Z}, +)$ is not isomorphic to $(\mathbb{Q}, +)$.

17. Is $(\mathbb{Q}, +)$ isomorphic to $(\mathbb{Q}^*, \times)$, where $\mathbb{Q}^*$ is the set of all nonzero rationals? Hint: Assume that there is an isomorphism $\theta : \mathbb{Q} \to \mathbb{Q}^*$.

Activity 9 - THE CIRCLE GROUP

Let $H = \{a + bi : a, b \in \mathbb{R}\}$ with $|a + bi| = \sqrt{a^2 + b^2} = 1$. Show that H is a subgroup of the multiplicative group of the nonzero complex numbers.

What do the cosets of the circle group look like? Explain why $H = \{\cos t + i \sin t : t \in \mathbb{R}\}$.

Let M be the set of 2×2 matrices of the form

$$m(x) = \begin{pmatrix} \cos x & \sin x \\ -\sin x & \cos x \end{pmatrix}, \text{ with } x \in \mathbb{R}$$

Prove that M is a multiplicative group.

Prove that M is isomorphic to the circle group, H.

What happens if you change H to

$$K = \{a + bi : a, b \in \mathbb{R}\} \text{ but with } |a + bi| \leq 1?$$

Direct Products

Known groups can be building blocks for forming new groups. If G_1 and G_2 are groups then the cartesian product $G_1 \times G_2$ is a group under the operation $(a,b) \circ (c,d) = (ac,bd)$ and is called the DIRECT PRODUCT. The identity is (e_1,e_2) and the inverse of (a,b) is (a^{-1},b^{-1}) since $(a,b) \circ (a^{-1},b^{-1}) = (e_1,e_2)$. Closure and associativity are easy to see.

Sometimes the operation is viewed as additive. For example, $\mathbb{Z}_2 \times \mathbb{Z}_3$ is an additive group. It has the six elements: (0, 0), (0, 1), (0, 2), (1, 0), (1, 1), (1, 2). the element (1, 1) is a generator: $2(1,1) = (1,1) + (1,1) = (0,2)$, and so on.

Theorem 1

If G_1 and G_2 are groups then $G_1 \times G_2$ is isomorphic to $G_2 \times G_1$.

Theorem 2

The direct product of abelian groups is also abelian.

The proofs for theorems 1 and 2 are left as problems.

Theorem 3: Fundamental Theorem of Finite Abelian Groups

Every finite abelian group can be written as a product of cyclic groups of prime power order. (No Proof).

Example 1

$\mathbb{Z}_6$ can be written as $\mathbb{Z}_2 \times \mathbb{Z}_3$; $\mathbb{Z}_{12}$ is isomorphic to $\mathbb{Z}_3 \times \mathbb{Z}_4$. The "penny" group can be expressed as $\mathbb{Z}_2 \times \mathbb{Z}_2$.

Example 2

$\mathbb{Z}_{18}$ can be written as $\mathbb{Z}_2 \times \mathbb{Z}_9$. Why do we exclude $\mathbb{Z}_3 \times \mathbb{Z}_6$ and $\mathbb{Z}_2 \times \mathbb{Z}_3 \times \mathbb{Z}_3$ as possibilities?

PROBLEMS - DIRECT PRODUCTS

1. Is $(1, 2)$ a generator of $\mathbb{Z}_2 \times \mathbb{Z}_3$?
2. Explain why $\mathbb{Z}_3 \times \mathbb{Z}_3$ is not cyclic.
3. Is $\mathbb{Z}_2 \times \mathbb{Z}_4$ cyclic? How about $\mathbb{Z}_3 \times \mathbb{Z}_4$?
4. What is the order of (2,6) in $\mathbb{Z}_4 \times \mathbb{Z}_{12}$?
5. What is the order of (3,10,9) in $\mathbb{Z}_4 \times \mathbb{Z}_{12} \times \mathbb{Z}_{15}$?
6. What are the orders of elements in $\mathbb{Z}_3 \times \mathbb{Z}_3 \times \mathbb{Z}_3$?
7. Can $\mathbb{Z}_2 \times \mathbb{Z}_8$ be isomorphic to $\mathbb{Z}_4 \times \mathbb{Z}_4$?
8. When is the group $\mathbb{Z}_m \times \mathbb{Z}_n$ isomorphic to $\mathbb{Z}_{mn}$?
9. How many abelian groups are there with order

   (a) 6 (b) 10 (c) 15 (d) 21 (e) 28 (f) pq, two primes?
10. Which direct product of cyclic groups is $\mathbb{Z}_{24}$ isomorphic to?
11. $G = \{1, 8, 12, 14, 18, 21, 27, 31, 34, 38, 44, 47, 51, 53, 57, 64\}$ is a group under multiplication modulo 65. What are the five possible direct products? And now, which one is it?
12. Find a subgroup of order 24 in $\mathbb{Z}_{30} \times \mathbb{Z}_{12}$.

13. Find a cyclic subgroup of $\mathbb{Z}_{30} \times \mathbb{Z}_{20}$ with order 15.

14. Write $\mathbb{Z}_2 \times \mathbb{Z}_2 \times \mathbb{Z}_3 \times \mathbb{Z}_5$ as a product $\mathbb{Z}_m \times \mathbb{Z}_n$.

15. Compute the order of 9 in $\mathbb{Z}_{30}$. Also, compute the order of 21. Why are these the same? What general theorem is this an example of?

16. Prove Theorem 1.

17. Prove Theorem 2.

Activity 10 - A HEISENBERG GROUP

Question: Can you have two nonisomorphic groups G_1 and G_2 such that the orders of elements in G_1 are exactly the same as the orders of elements in G_2?

Let

$$G = \mathbb{Z}_3 \times \mathbb{Z}_3 \times \mathbb{Z}_3 \text{ and let } H = \left\{ \begin{pmatrix} 1 & a & b \\ 0 & 1 & c \\ 0 & 0 & 1 \end{pmatrix} : a, b, c \in \mathbb{Z}_3 \right\},$$

be a multiplicative group of 3×3 matrices. What is the order of G? Of H? What is the order of each element in G? in H?

Is G isomorphic to H?

If you replace $\mathbb{Z}_3$ by $\mathbb{Z}_2$, what group of symmetries is H isomorphic to?

Permutation Groups

Each permutation of $1, 2, 3, 4$ is a one-to-one, onto function and can be viewed in several ways:

Table:

x	1	2	3	4
$f(x)$	2	4	3	1

Arrow: $1 \longrightarrow 2, 2 \longrightarrow 4, 3 \longrightarrow 3, 4 \longrightarrow 1$

Matrix: $\begin{pmatrix} 1 & 2 & 3 & 4 \\ 2 & 4 & 3 & 1 \end{pmatrix}$

Cycle: $(124); 1 \longrightarrow 2 \longrightarrow 4 \longrightarrow 1 \longrightarrow 2 \ldots$

With composition of functions as "multiplication", a group structure can be formed. For example, if $\alpha = \begin{pmatrix} 1 & 2 & 3 \\ 1 & 3 & 2 \end{pmatrix}$ and $\beta = \begin{pmatrix} 1 & 2 & 3 \\ 3 & 1 & 2 \end{pmatrix}$ the product $\alpha\beta$ becomes $\alpha\beta = \begin{pmatrix} 1 & 2 & 3 \\ 1 & 3 & 2 \end{pmatrix}\begin{pmatrix} 1 & 2 & 3 \\ 3 & 1 & 2 \end{pmatrix} = \begin{pmatrix} 1 & 2 & 3 \\ 3 & 2 & 1 \end{pmatrix} = (13)$, using both matrix and cycle forms. Also, $\beta\alpha = \begin{pmatrix} 1 & 2 & 3 \\ 2 & 1 & 3 \end{pmatrix} = (12)$, so composition of permutations is not commutative.

The set of all permutations on $X_n = \{1, 2, \ldots, n\}$ is denoted by S_n.

Definition. S_n is called the SYMMETRIC GROUP on $X_n = \{1, 2, 3, \ldots, n\}$.

Theorem 1

S_n is a group under composition of permutations.

Proof: To simplify notation and computation let's use $n = 3, 4, 5, 6$ for illustration purposes. The leap to general n just requires . . . in most cases.

Identity: $e = \begin{pmatrix} 1 & 2 & 3 & 4 \\ 1 & 2 & 3 & 4 \end{pmatrix}$ is easy to verify.

Inverses: The inverse of $\begin{pmatrix} 1 & 2 & 3 & 4 \\ a & b & c & d \end{pmatrix}$ is $\begin{pmatrix} a & b & c & d \\ 1 & 2 & 3 & 4 \end{pmatrix}$ since $\begin{pmatrix} 1 & 2 & 3 & 4 \\ a & b & c & d \end{pmatrix}\begin{pmatrix} a & b & c & d \\ 1 & 2 & 3 & 4 \end{pmatrix} = e.$

Closure: The composition of two permutations is another permutation.

Associativity: Easy to show.

Example 1

$\begin{pmatrix} 1 & 2 & 3 & 4 \\ 3 & 1 & 2 & 4 \end{pmatrix}^{-1} = \begin{pmatrix} 3 & 1 & 4 & 2 \\ 1 & 2 & 3 & 4 \end{pmatrix} = \begin{pmatrix} 1 & 2 & 3 & 4 \\ 2 & 4 & 1 & 3 \end{pmatrix}$. Moving columns around doesn't change the function.

Example 2

The permutation $\alpha = \begin{pmatrix} 1 & 2 & 3 & 4 & 5 & 6 \\ 6 & 4 & 1 & 2 & 5 & 3 \end{pmatrix}$ could be expressed in cycle form as $\alpha = (163)(24)(5)$. Note here that $(163) = (631) = (316)$, $(24) = (42)$, and (5) means leave 5 fixed, so we don't need to write it. Also, always start any cycle with the smallest integer - so we don't need (631) or (316). Just use (163).

Example 3

What is the order of $\alpha = (163)(24)$ in S_6?

The reader should verify the following computations:

$\alpha^2 = (163)(24)(163)(24) = (136)$

$\alpha^3 = \alpha^2\alpha = (136)(163)(24) = (24)$

$\alpha^4 = \alpha^2\alpha^2 = (163)$

$\alpha^5 = \alpha^2\alpha^3 = (136)(24) =$ well, we are nearly done. Notice that disjoint cycles commute.

$\alpha^6 = \alpha^3 \cdot \alpha^3 = (24)(24) = (1)$

So the order of $\alpha = (163)(24) = 6 = \text{lcm}(2, 3)$.

The "leap frog" technique is useful in computing powers of cycles. Let $\alpha = (14253)$.

$\alpha^2 = \dot{1}\ 4\ \dot{2}\ 5\ \dot{3}\ 1\ \dot{4}\ 2\ \dot{5}\ 3\ \dot{1}\ 4\ \dot{2}\ 5\ \dot{3} \ldots = (12345)$

$\alpha^3 = \dot{1}\ 4\ 2\ \dot{5}\ 3\ 1\ \dot{4}\ 2\ 5\ \dot{3}\ 1\ 4\ \dot{2}\ 5\ 3\ \dot{1}\ 4\ 2\ \dot{5}\ 3 \ldots = (15432)$.

Example 4

The order of (132) is 3, but the order of (123)(132) is 1 since (123)(132) = (1)(2)(3).

CAUTION: Each permutation has to be written as a product of DISJOINT CYCLES before you can compute its order.

PROBLEMS - PERMUTATION GROUPS

1. Express the inverse of $\begin{pmatrix} 1 & 2 & 3 & 4 \\ 3 & 4 & 1 & 2 \end{pmatrix}$ in cycle form.

2. Compute the order of each:

   (a) (12)(345) (b) (24) (c) (13)(12) (d) (132)(23) (e) (12)(13)(14)

   (f) (1234)(587)

3. Verify that (1423)=(14)(12)(13). Generalize.

4. Express (13)(12)(345) as a simple cycle.

5. Express $\alpha = (1234)(3456)$ as a product of disjoint cycles. What is the order of α?

6. Let $\alpha = (12)(34)$, $\beta = (13)(24)$, $\gamma = (14)(23)$. Compute $\alpha\beta$ and $\alpha\gamma$.

7. Express (132)(154)(123)(145) as a 3-cycle.

Groups of Symmetries

Symmetries of an Equilateral Triangle

Our goal here is to study such symmetries and to see how we can create group structures. Introduction of good notation will help. We start with the six symmetries of an equilateral triangle.

Lets take the following as our initial position, the identity, or "do nothing".

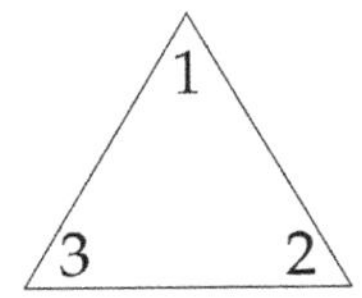

Let r denote a clockwise rotation through 120°, and so r^2 will be a 240° rotation:

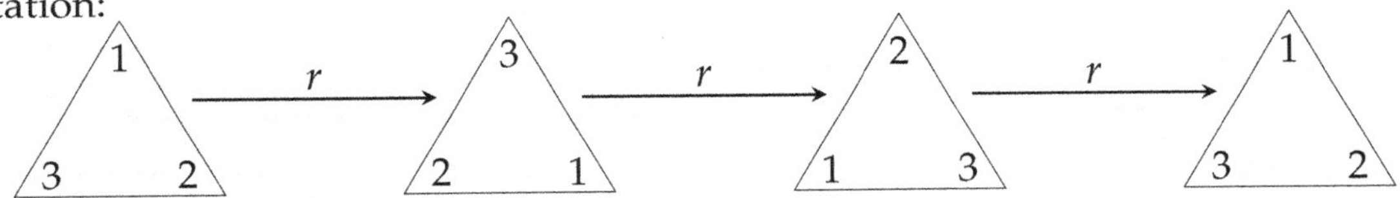

Applying r a third time will bring us back to our starting position, or $r^3 = 1$. $H = \{1, r, r^2\}$ is a nice subgroup.

Now do a flip, f, about the altitude holding vertex 1 fixed. This looks like:

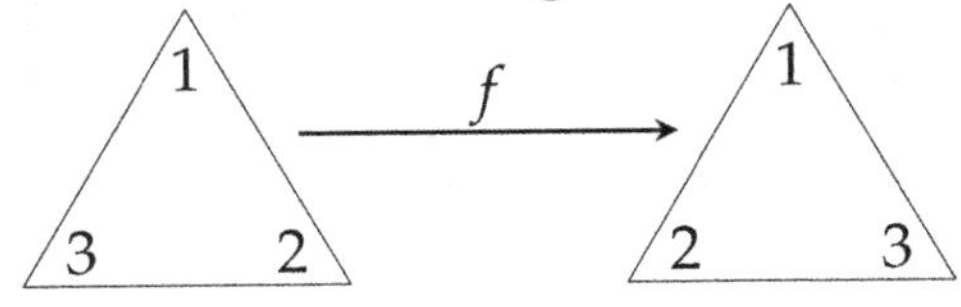

The reader should check that $r^2 f$ (r^2 followed by f, in that order) will yield another flip, one that pivots about vertex 3. The last symmetry, another flip, is given by rf which is a pivot about vertex 2.

Summary: The six symmetries of an equilateral triangle can be expressed in terms of just r and f. They are, $\{1, r, r^2, f, rf, r^2f\}$.

In the problems you are asked to make the 6×6 operation table for these six symmetries. As you try to complete the table you may run into trouble with a few like $fr, fr^2, frf\ldots$

Example 1

In the problems you are asked to draw a sequence of triangles to show $fr = r^2f$.

Example 2

$fr^2 = (fr)r = (r^2f)r = r^2r^2f = rf$

Example 3

$frf = (fr)f = r^2ff = r^2$.

There is something convenient about switching to cycle notation. Notice that $r = (132)$ and $f = (23)$. It is not hard to see that $r^2 = (123)$ and $rf = (13)$, $r^2f = (12)$. Now we can be independent of the pictures, and generalizations are easier to see.

BIG CAUTION! Rotate the pictures clockwise, but perform composition of functions (cycles) right to left just as you do when computing $f(g(x))$. Now cycle form corresponds with the $r - f$ version.

THE RECTANGLE GROUP

(OR HOW DOES THE POST OFFICE CANCEL STAMPS?)

There are four rigid motion symmetries of a rectangle.

I: DO NOTHING

1 2

4 3

H: HORIZONTAL FLIP

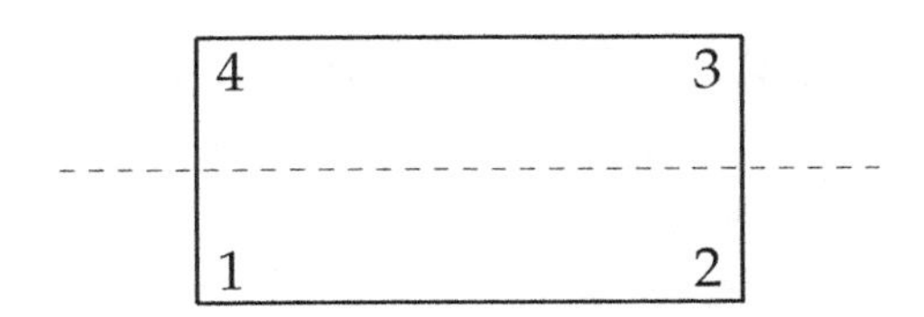

V: VERTICAL FLIP

2 1

3 4

R: 180° CLOCKWISE ROTATION

3 4

2 1

The interior numbers indicate their movement from I, the original position. The group table for $G = \{I, H, V, R\}$ is shown next:

$\circ$	I	H	V	R
I	I	H	V	R
H	H	I	R	V
V	V	R	I	H
R	R	V	H	I

G is evidently a Klein 4-group; the orders of $H, V,$ and R are 2.

The Rectangle Group in Cycle Form

It is easy to see that the following correspondence holds: $I = (1), H = (14)(23), V = (12)(34)$ and $R = (13)(24)$. In the problems you are asked to make the operation table for the rectangle group in cycle form.

PROBLEMS - GROUPS OF SYMMETRIES

1. Draw a sequence of triangles to show that $fr = r^2 f$.

2. Why is $f^2 = (1)$, the identity?

3. Show that $f(r^2 f) = r$.

4. Show that $(r^2 f)(rf) = r$.

5. Make the multiplication table for $\{1, r, r^2, f, rf, r^2 f\}$.

6. Make the multiplication table for $\{(1), (132), (123), (23), (13), (12)\}$.

7. Make the operation table for the rectangle group in cycle form.

Activity 11 - THE OCTIC GROUP

The octic group consists of the symmetries of a square: some rotations and flips about the diagonals and horizontal/vertical flips. Start with

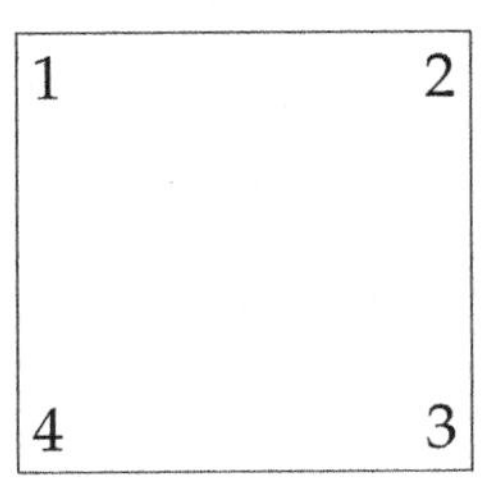

and let r be a 90° clockwise rotation, and let f denote a flip about the diagonal fixing 1 and 3. Determine that there are 8 symmetries. The easiest are e, r, r^2, r^3. Show that $fr = r^3 f$ and use this to complete the 8×8 operation table for the octic group $\{e, r, r^2, r^3, f, rf, r^2 f, r^3 f\}$.

Activity 12 - A TETRAHEDRON

List the 12 symmetries of a regular tetrahedron in cycle form. For example, spinning about vertex 1 gives two : (234) and (243). What is the other type?

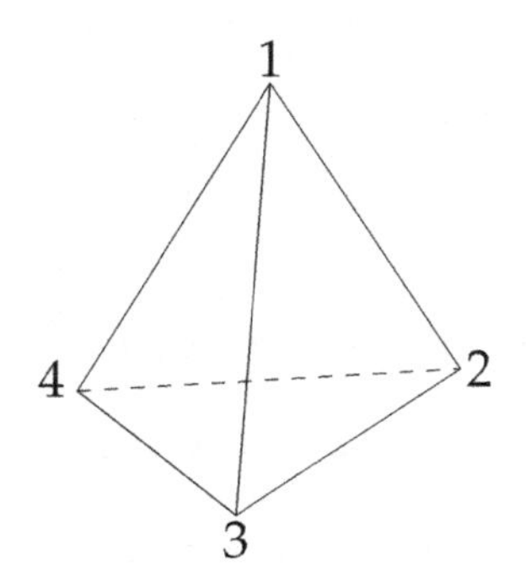

Quotient Groups

Cosets and Normal Subgroups

Lagrange's Theorem states that the order of any subgroup H divides the order of a (finite) group G. This is (was) proved by partitioning the elements of G into nonoverlaping sets called cosets. For $a \in G$ a (left) coset of H is the set $\{aH\}$ consisting of all products ah with $h \in H$. The number of cosets is called the index of H in G. A right coset is the set $\{Ha : a \in G\}$. If $aH = Ha$ for every $a \in G$, as sets, we say that H is a normal subgroup. For an additive group, the notation $a + H = H + a$ is used.

Example 1

Let $S_3 = \{1, r, r^2, f, rf, r^2f\}$ be the rigid transformations of an equilateral triangle. $H = \{1, f\}$ is not a normal subgroup: $rH = \{r, rf\}$, but $Hr = \{r, fr\} = \{r, r^2f\}$. As sets, $rH \neq Hr$.

Example 2

$H = \{1, r, r^2\}$ is a normal subgroup of $G = S_3$. $aH = Ha$ if $a \in H$. So now try the other three: $fH = \{f, r^2f, rf\} = Hf$, $rfH = \{rf, f, r^2f\} = Hrf$, and also $r^2fH = Hr^2f$. So $aH = Ha$, for all $a \in G$.

The good news is that if G is abelian, all subgroups are normal. Here is another helpful result: If the index of H in G is 2, H is normal. That's why it is easy to see that the H in example 2 is normal.

It turns out that if H is normal you can multiply cosets! You get the following nice process: $aHbH = abH$. This is not hard to show: $abH \subseteq aHbH$ is easy to show. For $aHbH \subseteq abH$, use normality.

Theorem 1

Let N be a normal subgroup of G. The product $aNbN$ of cosets is the coset abN.

Theorem 2

The (left) cosets of a normal subgroup H in G form a group.

Proof:

1. Closure: $aHbH = abH$
2. Identity: If e is the identity in G, eH is the identity in the new group since $eHaH = eaH = aH = H = aeH = aHeH$.
3. Inverses: The inverse of aH is $a^{-1}H$ since $aHa^{-1}H = a^{-1}aH = H = a^{-1}HaH$.
4. Associativity: $aH(bHcH) = aHbcH = abcH$, and $(aHbH)cH = abHcH = abcH$.

Definition. If N is a normal subgroup of G, the set of all left cosets aN form a group called the QUOTIENT GROUP, or Factor Group, denoted G/N.

Example 3

Let $G = \mathbb{Z}_6$ and $H = \{0, 3\}$. Since $\mathbb{Z}_6$ is abelian, H is a normal subgroup. The three cosets are H, $1 + H = \{1, 4\}$ and $2 + H = \{2, 5\}$ and $G/H = \{H, 1 + H, 2 + H\}$. We next make the addition table for $\mathbb{Z}_6$, with the elements rearranged as $0, 3, 1, 4, 2, 5$; next to it make the operation table for G/H:

$\oplus$	0	3	1	4	2	5
0	0	3	1	4	2	5
3	3	0	4	1	5	2
1	1	4	2	5	3	0
4	4	1	5	2	0	3
2	2	5	3	0	4	1
5	5	2	0	3	1	4

	H	$1+H$	$2+H$
H	H	$1+H$	$2+H$
$1+H$	$1+H$	$2+H$	H
$2+H$	$2+H$	H	$1+H$

These two tables look almost the same!

There are several features of this example to notice:

1. The three cosets partition the elements of $\mathbb{Z}_6$ into three disjoint sets.
2. $H = \{0, 3\}$ is the identity coset.
3. Some funny addition is going on: $(1 + H) + (2 + H) = \{1, 4\} + \{2, 5\}$ $= \{1 + 2, 1 + 5, 4 + 2, 4 + 5\} = \{3, 6, 6, 9\} = \{0, 3\} = H$.
4. The inverse of $1 + H$ is $2 + H$.
5. The index of H in G is 3.

Example 4

Let $G = V_{15} = \{1, 2, 4, 7, 8, 11, 13, 14\}$ and $H = \{1, 11\}$. H is normal in G (why?). The four cosets of H are H, $2H = \{2, 7\}$, $4H = \{4, 14\}$, and $8H = \{8, 13\}$. The quotient group table is:

	H	$2H$	$4H$	$8H$
H	H	$2H$	$4H$	$8H$
$2H$	$2H$	$4H$	$8H$	H
$4H$	$4H$	$8H$	H	$2H$
$8H$	$8H$	H	$2H$	$4H$

Which group of order 4 is this? $\mathbb{Z}_4$ or the Klein 4-group?

Example 5

Let $G = \mathbb{Z}$, and $H = 3\mathbb{Z}$. Then the cosets in G/H are $0+H$, $1+H$ and $2+H$. If $H = 2\mathbb{Z}$ the cosets are simply the set of even integers, and the set of odd integers.

PROBLEMS - QUOTIENT GROUPS

1. List the left cosets of the subgroup $3\mathbb{Z}$ in $\mathbb{Z}$. Now list the right cosets. These are the same since the additive group $\mathbb{Z}$ is an abelian group.

2. List the left cosets of $H = \{1, rf\}$ in S_3. Now list the right cosets. What do you notice?

3. Let $H = \{..., -10, -5, 0, 5, 10, ...\}$. Find all the left cosets of H in $\mathbb{Z}$. Are $-3 + H$ and $-8 + H$ in the same coset? Where would you find 21? How about -17?

4. Let $H = \{1, 8\}$ be a subgroup of $V_9 = \{1, 2, 4, 5, 7, 8\}$, the multiplicative group of the invertibles mod 9. Make all left cosets of H in V_9, and make the group table for V_9/H.

5. $H = \{1, r, r^2\}$ is a normal subgroup of $S_3 = \{1, r, r^2, f, rf, r^2f\}$. Make the group table for the quotient group S_3/H.

6. What is the size of $3 + H$ where $H = \{0, 6, 12\}$ is a subgroup of $\mathbb{Z}_{18}$? What is the order of $3 + H$ in $\mathbb{Z}_{18}/H$? Is this quotient group cyclic?

7. If G is an abelian group, explain why every subgroup of G is normal.

8. Determine the elements of the quotient group for each of the following:

   (a) $G = \mathbb{Z}_{12}$ $H = \{0, 4, 8\}$

   (b) $G = \mathbb{Z}$ $H = 2\mathbb{Z}$

   (c) $G = D_4$ $H = \{1, r, r^2, r^3\}$ where D_4 is the group formed from the symmetries of a square

   (d) $G = V_{15}$ $H = \{1, 4, 11, 14\}$

   (e) $G = V_{15}$ $H = \{1, 4\}$

9. Let $G = [a]$ be a cyclic group of order 21 generated by a, and let H be a subgroup having index 3. List the elements of H and the elements of G/H. Make the operation table for the quotient group G/H.

10. Let G be a cyclic group of order 91 and H be a subgroup having index 7. List the cosets of the quotient group G/H.

11. If the index of a subgroup H in G is 2, prove that H is normal.

12. The six roots of $x^6 - 1 = 0$ form a multiplicative group $G = \{1, r, s, -1, -r, -s\}$ where 1, r, s are roots of $x^3 - 1 = 0$. Form the left cosets of $H = \{1, r, s\}$ and make the operation table for G/H.

13. Let $G = \{000, 001, 010, 011, 100, 101, 110, 111\}$ under bitwise addition mod 2, and $H = \{000, 011\}$. List the left cosets of H.

14. The elements of the Quaternion group G are $\{1, -1, i, -i, j, -j, k, -k\}$. Find a normal subgroup H and make the operation table for G/H.

15. Prove that if G is an abelian group, so is the quotient group G/H for any normal subgroup H.

16. Let $G = \mathbb{Z}_4 \times \mathbb{Z}_4$ and let N be the cyclic subgroup generated by the element $(3, 2)$. Show that G/N is isomorphic to $\mathbb{Z}_4$.

17. $\mathbb{Q}/\mathbb{Z}$ is an additive abelian group, with infinite order. What is the order of the coset $2/7 + \mathbb{Z}$?

18. $H = \{(x, 5x) : x \in \mathbb{R}\}$ is a subgroup of the additive group $\mathbb{R} \times \mathbb{R}$. Give a geometrical description of H and of the coset $(2, 7) + H$. What do the cosets of H look like? What do the cosets of the circle group in the complex numbers look like?

19. Let $N = \{(x, y) : y = -x\}$ be a subgroup of the additive group $\mathbb{R} \times \mathbb{R}$. Describe the cosets of N.

A Brief Look at Rings

This type of structure should ring a bell. As in the integers $\mathbb{Z}$, in a ring we can add, subtract, multiply and even distribute. So think of $\mathbb{Z}$ as our model of a ring.

Definition. A ring R is a set with two binary operations such that

(a) R is an abelian group under addition.

(b) R is closed and associative under multiplication.

(c) Multiplication is distributive over addition, ie, $a(b+c) = ab + ac$ and $(b+c)a = ba + ca$.

A commutative ring is a ring where $ab = ba$.

Example 1

$\mathbb{Z}, \mathbb{Q}, \mathbb{R},$ and $\mathbb{C}$ are commutative rings.

Example 2

$\mathbb{Z}_m$, the ring of integers modulo m, is commutative.

Example 3

The set $\mathbb{Z}[x]$ of all polynomials in x with coefficients in $\mathbb{Z}$ is a commutative ring.

Definition. If a ring R has a multiplicative identity 1, then an element a in R is an invertible if there is an a^{-1} such that $a\,a^{-1} = 1$; 1 is called the unity.

Example 4

$2\mathbb{Z}$ is a commutative ring with no unity.

Example 5

$(\mathbb{Z}, \oplus, \otimes)$ is a ring where $a \oplus b = a + b - 1$ and $a \otimes b = a + b - ab$. Closure under $\oplus$ is clear since $a \oplus b = a + b - 1 \in \mathbb{Z}$. The additive identity is 1 since $a \oplus 1 = a + 1 - 1 = a = 1 \oplus a$. The inverse of a is $2 - a$ since $a \oplus (2 - a) = a + 2 - a - 1 = 1$. The binary operation $\oplus$ is associative since: $a \oplus (b \oplus c) = a \oplus (b + c - 1) = a + b + c - 2$ and $(a \oplus b) \oplus c = (a + b - 1) \oplus c = a + b - 1 + c - 1 = a + b + c - 2$.

Closure under multiplication is clear since $a \otimes b = a + b - ab \in \mathbb{Z}$. Finally, we check distributivity: $a \otimes (b \oplus c) = a \otimes (b + c - 1) = a + b + c - 1 - (ab + ac - a)$ and $(a \otimes b) \oplus (a \otimes c) = (a + b - ab) \oplus (a + c - ac) = 2a + b + c - ab - ac - 1$. The reader should check that $\otimes$ is associative.

A subring (like a subgroup) of a ring R is a subset of R that is a ring. The set $\{0, \pm 5, \pm 10, \dots\}$ is a subring of $\mathbb{Z}$.

PROBLEMS - A BRIEF LOOK AT RINGS

1. Show that $\mathbb{Z}[\sqrt{2}] = \{a + b\sqrt{2} : a, b \in \mathbb{Z}\}$is a ring.
2. In a ring R, show that $a^2 - b^2 = (a + b)(a - b)$ if and only if R is commutative.
3. Show that $\mathbb{Z}[i] = \{a + bi : a, b \in \mathbb{Z}\}$ is a ring. $\mathbb{Z}[i]$ is called the ring of Gaussian Integers.
4. Find all invertibles in $\mathbb{Z}[i]$.
5. Find the unity in $S = \{0, 2, 4, 6, 8\}$ under addition mod 10.

6. What are the invertibles in $\mathbb{Z} \times \mathbb{Z}$?

7. Let $R = \{0, 1, c\}$ be a ring with unity.

   (a) Show that $1 + 1 = c$ and that $1 + 1 + 1 = 0$.

   (b) Show that $c^2 = 1$.

   (c) Make the $\times$ and $+$ tables for R.

8. Let $R = \{0, 1, c, d\}$ be a ring (with unity 1) with c, d invertibles. Make the multiplication table for R.

9. For $a, b \in \mathbb{Q}$, define $\oplus$ and $\otimes$ by $a \oplus b = ab$ and $a \otimes b = a + b$. Is $(\mathbb{Q}, \oplus, \otimes)$ a ring?

Activity 13 - COMPLETE THE RING

Complete the multiplication table for the ring $R = \{a, b, c, d\}$

+	a	b	c	d
a	a	b	c	d
b	b	a	d	c
c	c	d	a	b
d	d	c	b	a

*	a	b	c	d
a	a	a	a	a
b	a	b		
c	a			a
d	a	b	c	

Integral Domains

When you were asked to solve $x^2 - 7x + 12 = 0$, you set each factor in $(x-3)(x-4)$ to zero and solved $x - 3 = 0$ or $x - 4 = 0$. Now try that in $\mathbb{Z}_{12}$; you get 3, 4 and 7 as solutions. This quadratic has three roots.

Definition. If a and b are nonzero elements of a ring and $ab = 0$, we call a and b 0-divisors.

Example 1

n $\mathbb{Z}_{12}$, the 0-divisors are $2, 3, 4, 6, 8, 9, 10$ since $2 \cdot 6 = 3 \cdot 4 = 8 \cdot 9 = 6 \cdot 10 = 0$ in $\mathbb{Z}_{12}$. These seven elements are precisely those numbers not relatively prime to 12.

Definition. An integral domain is a commutative ring D with a unity, that has no 0-divisors.

These are	These are not
$\mathbb{Z}, \mathbb{Q}, \mathbb{R}$	$\mathbb{Z}_6, \mathbb{Z}_{12}$
$\mathbb{Z}_p, p$ a prime	$\mathbb{Z}_m, m$ composite
$\mathbb{Z}[\sqrt{2}]$	$\mathbb{Z} \times \mathbb{Z}$
$\mathbb{Z}[x]$	$2\mathbb{Z}$
$\mathbb{Z}[i]$	$\mathbb{Z}_5[i]$

PROBLEMS - INTEGRAL DOMAINS

1. List all 0-divisors in $\mathbb{Z}_{20}$. What are the invertibles?

2. Find all solutions to $x^2 - 4x + 3 = 0$

   (a) in $\mathbb{Z}_{12}$

   (b) in $\mathbb{Z}_{11}$

3. Find a 0-divisor in $\mathbb{Z}_5[i] = \{a + bi : a, b \in \mathbb{Z}_5\}$.

4. Show that $\mathbb{Z}\times\mathbb{Z}$, with multiplication and addition defined coordinatewise, is not an integral domain.

5. Why is $2\mathbb{Z}$ not an integral domain?

6. Let $S = \{a, b, c\}$ and $P(S)$ be the power set of S, ie, the set of all subsets of S including ϕ and S. Define the product AB to be $A \cap B$ and the sum $A + B$ to be $(A \cup B) - (A \cap B)$, the elements in $A \cup B$ but not those in $A \cap B$.

   (a) Show that $P(S)$ is a commutative ring.

   (b) What is the unity?

   (c) What acts like a "0"?

   (d) Is $P(S)$ an integral domain?

7. In Z_6 show that $ab = ac$ does not imply $b = c$.

8. Show that $M_2(\mathbb{Z}_2)$, the set of all 2×2 matrices with entries in $\mathbb{Z}_2$ with the usual matrix operations is not an integral domain.

9. Prove that right and left cancellation laws hold in a ring R if and only if R has no zero divisors. In other words, if $c \neq 0$ and c is not a 0-divisor, then either $ac = bc$ or $ca = cb$ implies $a = b$.

10. Is the ring of Gaussian integers an integral domain?

11. Is the direct product of integral domains an integral domain?

12. Suppose $D = \{0, d_2, d_3, \ldots, d_n\}$ is an integral domain. Prove that $\{d_2, d_3, \ldots, d_n\}$ is a multiplicative group.

FIELDS – THE FINALE

A field is a commutative ring with unity where every nonzero element is an invertible. In other words, a field is an integral domain whose nonzero elements form a multiplicative group. More formally, a field F

(a) is an abelian group under addition

(b) is an abelian group under multiplication (don't count 0)

(c) has the property that multiplication is distributive over addition.

Example 1

$\mathbb{Q}$, $\mathbb{R}$, $\mathbb{C}$, $\mathbb{Z}_p$, $\mathbb{Q}[\sqrt{2}]$ are fields.

Example 2

$\mathbb{Z}_3[i]$, $\mathbb{Z}_7[i]$ and $\mathbb{Z}_{11}[i]$ are fields; but $\mathbb{Z}_2[i]$, $\mathbb{Z}_5[i]$ and $\mathbb{Z}_{13}[i]$ are not. [Note that when $p \equiv 3 \pmod 4$, $\mathbb{Z}_p[i]$ is a field].

Theorem 1

Every finite integral domain D is a field.

Proof: Let $a \in D$ and $a \neq 0$; we show that a has a multiplicative inverse. Let $D = \{0, 1, a_1, a_2, \ldots, a_n\}$. The elements $a \cdot 1, a\,a_1, a\,a_2, \ldots, a\,a_n$ are surely distinct since D is an integral domain, and none of these products is 0 since D has no zero-divisors. So one of these must be 1; let's say $a\,a_i = 1$. But then a_i is the inverse of a.

PROBLEMS - FIELDS

1. Verify that $\mathbb{Q}[\sqrt{2}]$ is a field. Show that $2 + 3\sqrt{2}$ has an inverse.

2. Why is $\mathbb{Z}_2[i]$ not a field?

3. Show that each of the following is not a field by finding 0-divisors:

   (a) $\mathbb{Z}_{13}[i]$. (b) $\mathbb{Z}_{17}[i]$.

4. Is $\mathbb{Z}[\sqrt{2}]$ a field?

5. Let $F = \{0, 2, 4, 6, 8\}$ under addition and multiplication modulo 10. Prove that F is a field.

6. $\mathbb{Z}_3[i]$ is a field with 9 elements.

   (a) Make the 8 by 8 multiplication table.

   (b) Make a table of inverses and orders of each element.

   (c) Which familiar group is the multiplication group isomorphic to?

Selected Answers

1. Binary Operations

1. **(a)** No; $a - b \neq b - a$. **(b)** No. **(c)**Yes. **(d)** Yes; $1 \cdot a = a$, No inverses.

3. **(a)** Yes; $A \cup B = B \cup A$ **(b)** No; $\emptyset$ is the identity; no set in X satisfies $\{a\} \cup X = \emptyset$.

5. **(a)** $2 \circ 5 = 175$; $3 \circ 2 = 625$ **(b)** No; no identity.

7. Easy.

9. $e = 10$; no inverses.

11. **(a)** Yes. **(b)** $e = 1$ since $1 \square a = 1 + a - 1 = a = a \square 1$. **(c)** Yes; the inverse of m is $2 - m$ since $m \square (2 - m) = 2 - 1 = 1$.

13. **(a)** No; $1 \square 2 = 6$, but $2 \square 1 = 9$. **(b)** Three, you should find these.

15. Try $1 \square 0 \square 2$ and then $0 \square 1$ for commutativity.

17. **(a)** No; $2^3 \neq 3^2$. **(b)**No, try $2 \square 1 \square 3$. **(c)**$a^{(b^c)}$.

2. Closure

1. **(a)** No; $1 + 4 = 5$. **(b)**No; $4 - 1 = 3$. **(c)** Yes, $a^2b^2 = (ab)^2$

3. No; $4 + 6 = 10 \notin C$.

5. **(a)** No; $1 + 4 = 5$, **(b)** Yes; $(3h + 1)(3k + 1) = 9hk + 3(h + k) + 1$.

7. $H = \{0, \pm 6, \pm 12, \dots\}$; $G \cap H = \{0, \pm 12, \pm 24, \dots\}$, $12h - 12k = 12(h - k)$.

9. Yes; yes.

11. No

13. Let S be closed under subtraction. If $a \in S, a - a = 0 \in S$.

15. **(a)** $5 = 1 + 4,\ 13 = 4 + 9,\ 65 = 1 + 64$. **(b)** Yes, we can derive that

$$(m_1^2 + n_1^2)(m_2^2 + n_2^2) = (m_1m_2 - n_1n_2)^2 + (n_1m_2 + m_1n_2)^2$$

17. Yes

19. 7 and 9

3. Groups

1. Check the four axioms.

3. The identity is 16.

5. $7^1 = 7, 7^2 = 9, 7^3 = 3, 7^4 = 1; 9^1 = 9, 9^2 = 1$

7. G is cyclic and generated by i or $-i$.

9. $G = \{00, 01, 10, 11\}$ under bitwise addition $\mod 2$.

11. Mimic the table for B^2; G is Abelian, but not cyclic.

13. 0 is the additive identity; the inverse of $7a$ is $-7a$.

15. $\{0\}, \{0, 3\}, \{0, 2, 4\}$ and $\mathbb{Z}_6$ are the subgroups of $\mathbb{Z}_6$.

17. $\{0\}, \{0, 6\}, \{0, 3, 6, 9\}, \{0, 4, 8\}, \{0, 2, 4, 6, 8, 10\}$, and $\mathbb{Z}_{12}$ are the subgroups of $\mathbb{Z}_{12}$.

19. The elements 0 and 2 have no inverses.

21. $1, 2, 3, 4$ are generators. For example, $0 \cdot 2 = 0$, $1 \cdot 2 = 2, 2+2 = 4$, $2+2+2 = 1$, $2 + 2 + 2 + 2 = 3$.

23. **(a)** $0, 3$. **(b)** $0, 4$. **(c)** $0, 2, 3, 5$.

25. The identity is $a^0 = 1$; The inverse of a^h is a^{-h}. Since $a^h a^k = a^{h+k}$, H is closed.

27. The identity is $\begin{pmatrix} 0 & 0 \\ 0 & 0 \end{pmatrix}$; the inverse of $\begin{pmatrix} a & b \\ c & d \end{pmatrix} = \begin{pmatrix} -a & -b \\ -c & -d \end{pmatrix}$.

29. $0x + 0$ is the identity; the inverse of $ax + b$ is $-ax - b$.

31. $(ab)^{k+1} = (ab)(ab)^k = (ab)(a^k b^k) = (ba)(a^k b^k) = ba^{k+1} b^k = a^{k+1} b^{k+1}$ since b commutes with a.

4. More on Cyclic Groups

1. V_9 is cyclic with 2 as a generator.

3. $H = \{e, a^8, a^{16}\}$; $o(a^5) = 24$. List $(a^5)^k$ for $k = 0, 1, \ldots, 24$.

5. $\{1, 3, 7, 9\}$ is cyclic. $G = [3]$.

7. 6; $[25] = \{0, 25, 20, 15, 10, 5\}$

9. **(a)** 6. **(b)** 6. **(c)** 8. **(d)** 16.

11. **(a)** ω is a root of $\omega^3 - 1 = (\omega - 1)(\omega^2 + \omega + 1)$. **(b)** $\omega^3 - 1 = 0$. **(c)** 22. **(d)** 8, since $(\cos \frac{\pi}{4} + i \sin \frac{\pi}{4})^8 = 1$ and no lower power yields 1. **(e)** ∞; start taking powers using De Moivre's Theorem.

13. **(a)** 3. **(b)** 2.

Wilson's Theorem

1. $2 \cdot 7 \equiv 3 \cdot 9 \equiv 4 \cdot 10 \equiv 5 \cdot 8 \equiv 6 \cdot 1$ leaving $12 \equiv -1(13)$.

3. $2 \cdot 3 \cdot 4 \cdot \cdots \cdot p-2$ has an even number of elements x that pair up with x^{-1}. Now use $x \cdot x^{-1}$.

5. Use $(p-2)! \equiv 1(p)$.

6. $x = 1$ or $x = -1$. If $x^2 - 1 = (x-1)(x+1) \equiv 0(p)$, then $x = 1$ or $x = -1$. Since $(x+1) - (x-1) = 2$, either $p|(x+1)$ or $p|(x-1)$, but not both.

7. Hints: $30 \equiv -29(59)$ and $31 \equiv -30(61)$.

5. Lagrange's Theorem

1. $H = \{0,6\}, 1+H = \{1,7\}, 2+H = \{2,8\}, 3+H = \{3,9\}, 4+H = \{4,10\}$, and $5+H = \{5,11\}$.

3. $V_9 = \{1,2,4,5,7,8\}; H = \{1,8\}, 2H = \{2,7\}, 4H = \{4,5\}$. $\{1,4,7\}$ is another subgroup.

5. If $o(G) = p$, then by Lagrange's Theorem every element $g \neq e$ has order p. So any one of these is a generator.

7. $1,3,5,9,15,45$.

9. $H = \{e, a^{13}, a^{26}, \ldots, a^{78}\}$ has order 7; hence the index is 13.

6. Isomorphisms

1. $\theta(x) = 2x$.

3. For onto, choose $b \in \mathbb{R}^+$, then solve $2^x = b$ for $x = \log_2 b$. For one-to-one, if $a \neq b$ in $\mathbb{R}, 2^a \neq 2^b$ in $\mathbb{R}^+$.

5. Try $\theta(x) = \ln x$.

7. $\theta(e)\theta(e) = \theta(ee) = \theta(e) = e'\theta(e)$. Right cancellation gives $\theta(e) = e'$. $xx^{-1} = e$ gives $\theta(x)\theta(x^{-1}) = \theta(e) = e'$ and so $\theta(x^{-1}) = [\theta(x)]^{-1}$.

9. Let $\theta(a) = a'$ and let $o(a) = q$ and $o(a') = q'$, then $e = a^q$ and $\theta(a^q) = [\theta(a)]^q$. So $[\theta(a)]^q = e'$. Since q' is the least integer m with $[\theta(a)]^m = e'$, we have $q \geq q'$. Similarly, $q' \geq q$ and hence $q = q'$.

11. Since $\sqrt{xy} = \sqrt{x}\sqrt{y}$, $\theta(xy) = \theta(x)\theta(y)$. θ is one-to-one and onto.

13. $V_{12} = \{1,5,7,11\}$. $\theta(1) = 1, \theta(3) = 5, \theta(5) = 7, \theta(7) = 11$ is an isomorphism. Alternatively, they are both Klein 4-groups.

15. -1 has order 2 in $(\mathbb{R}^*, \times)$, but $(\mathbb{R}^+, \times)$ has no element of order 2, since $x^2 = 1 \implies x = 1$.

17. Let $a \in \mathbb{Q}$ and suppose $\theta(a) = -1$; -1 must have a preimage if θ is an isomorphism. Then $-1 = \theta(\frac{a}{2} + \frac{a}{2}) = [\theta(\frac{a}{2})]^2$, a contradiction.

7. Direct Products

1. Yes; you should compute $m(1,2)$ for $m = 0, 1, \ldots, 5$.

3. No; Yes. $(1,1)$ is a generator for $\mathbb{Z}_3 \times \mathbb{Z}_4$. In fact, $\mathbb{Z}_3 \times \mathbb{Z}_4$ is isomorphic to $\mathbb{Z}_{12}$, a cyclic group.

5. 60; $o(3)$ in $\mathbb{Z}_4$ is 4; $o(10)$ in $\mathbb{Z}_{12}$ is 6; $o(9)$ in $\mathbb{Z}_{15}$ is 5. LCM$(4,6,5) = 60$.

7. No; compare orders

9. **(a)** 1. **(b)** 1. **(c)** 1. **(d)** 1. **(e)** 1.

11. $\mathbb{Z}_{16}, \mathbb{Z}_4 \times \mathbb{Z}_4, \mathbb{Z}_2 \times \mathbb{Z}_8, \mathbb{Z}_2 \times \mathbb{Z}_2 \times \mathbb{Z}_4, \mathbb{Z}_2 \times \mathbb{Z}_2 \times \mathbb{Z}_2 \times \mathbb{Z}_2$. You should compare orders to see which one it is.

13. $\mathbb{Z}_3 \times \mathbb{Z}_5$ is one possibility. How about $[(2,0)]$ or $[(10,4)]$. There are others.

15. $o(9) = 10; o(21) = 10$. 9 is the inverse of 21 since $9 + 21 \equiv 0$. In general, $o(x) = o(x^{-1})$.

17. Let G_1 and G_2 be Abelian groups. Choose (a,b) and $(c,d) \in G_1 \times G_2$. Then $(a,b)(c,d) = (ac, bd) = (ca, db) = (c,d)(a,b)$.

8. Permutation Groups

1. (13)(24)

3. $(1abcd) = (1a)(1b)(1c)(1d)$ and similarly for an $n-$cycle.

5. (124)(356)

7. (142)

9. Symmetries of an Equilateral Triangle

2. A flip, followed by a flip, brings you back to the starting point.

3. $fr^2f = (fr)(rf) = r^2frf = r^2r^2ff = r^4f^2 = r$.

10. Quotient Groups

1. $3\mathbb{Z}, 1 + 3\mathbb{Z}, 2 + 3\mathbb{Z}; 3\mathbb{Z}, 3\mathbb{Z} + 1, 3\mathbb{Z} + 2$.

3. $H, 1 + H, 2 + H, 3 + H, 4 + H$; Yes; $21 \in 1 + H, -17 \in 3 + H$.

5. $H = \{1, r, r^2\}$ and $fH = \{f, r^2f, rf\}$ are the only cosets.

7. $aH = Ha$ for all $a \in G$ and every subgroup H.

9. $H = \{e, a^3, a^6, a^9, a^{12}, a^{15}, a^{18}\}; G/H = \{H, aH, a^2H\}$

11. If the index is 2, there are only two cosets, say H and aH. But then Ha must be either H or aH, so $Ha = Ha$ since $Ha \neq H$.

13. $001 + H = \{001, 010\}, 100 + H = \{100, 111\}, 101 + H = \{101, 110\}$.

15. $aH \cdot bH = abH = baH = bHaH$

17. Order of $\frac{2}{7} + \mathbb{Z}$ is 7 since $7(\frac{2}{7} + \mathbb{Z}) = 2 + \mathbb{Z} = \mathbb{Z}$.

19. Lines parallel to $y = -x$.

Integral Domains

1. 2, 4, 5, 6, 8, 10, 12, 14, 15, 16, 18. The invertibles are 1, 3, 7, 9, 11, 13, 17, 19.

3. $(1 + 2i)(1 + 3i) = 1 + 5i - 6 = 0$; or $(2 + i)(2 - i) = 0$.

5. No unity.

7. $2 \cdot 4 = 2 \cdot 1$, but $4 \neq 1$.

9. If R has no 0–divisors, then $ac = bc$ implies $ac - bc = (a - b)c = 0$ and so $a - b = 0$ since $c \neq 0$. You should now prove the converse.

11. No. $(0, 1)(1, 0) = (0, 0)$.

11. A Brief Look at Rings

1. Check the three ring properties. For example, $(a + b\sqrt{2})(c + d\sqrt{2}) = ac + (bc + ad)\sqrt{2} + 2bd$ shows closure under multiplication.

3. Check Properties. $(a + bi)(c + di) = ac - bd + (bc + ad)i$ shows closure under multiplication.

5. 6 is the unity.

7. **(a)** First make the addition table, then $1 + 1 = c, 1 + 1 + 1 = c + 1 = 0$. **(b)** $c^2 = (1 + 1)(1 + 1) = 1 + 1 + 1 + 1 = 0 + 1 = 1$.

9. Distributivity fails. $1 \otimes (2 \oplus 3) = 1 \otimes 6 = 7$. But $(1 \otimes 2) \oplus (1 \otimes 3) = 3 \oplus 4 = 12$.

12. Fields

1. $\frac{1}{2+3\sqrt{2}} = \frac{1}{2+3\sqrt{2}} \cdot \frac{2-3\sqrt{2}}{2-3\sqrt{2}} = \frac{2-3\sqrt{2}}{-14} = -\frac{1}{7} + \frac{3}{14}\sqrt{2}$

3. **(a)** $(2 + 3i)(2 - 3i) = 4 + 9 = 0$ **(b)**$(4 + i)(4 - i) = 0$

5. 6 is the unity; $2 \cdot 8 \equiv 4 \cdot 4 \equiv 6 \cdot 6 \equiv 8 \cdot 2 \equiv 6$ provides inverses.

Primer Index

AN INTERACTIVE WORKBOOK ON

Abstract Algebra

and some number theory

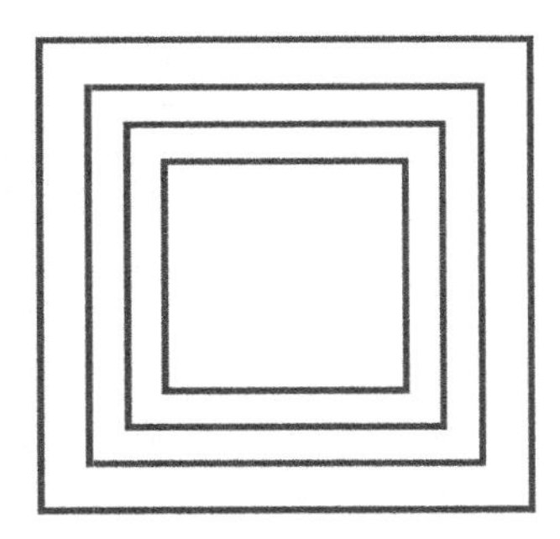

Richard Grassl
University of Northern Colorado
Emeritus Professor of
Mathematical Sciences
richard.grassl@unco.edu

Tabitha Mingus
Associate Professor, Collegiate
Mathematics Education
Western Michigan University
tabitha.mingus@umich.edu

Edited and typeset in LaTeX by Michael K. Petrie

Contents

DAY 1 PROPERTIES OF THREE TABLES

$\bullet$ = usual complex multiplication

$\bullet$	1	-1	i	$-i$
1				
-1				
i				
$-i$				

PROPERTIES, OBSERVATIONS

1.

2.

3.

$\otimes$ = units digit in regular multiplication

$\otimes$	1	3	7	9
1				
3				
7				
9				

1.

2.

3.

$\oplus$ = bitwise addition, 0 if same, 1 if different

$\oplus$	00	01	10	11
00				
01				
10				
11				

1.

2.

3.

WORKSHEET ON CLOSURE I

A set $S = \{a, b, c, \dots\}$ is closed under a binary operation $\circ$ if whenever x and y are elements of S so is $x \circ y$.

For each of the following if the answer is yes, give a reason and if no, provide a counterexample.

Task 1 Is $E = \{0, 2, 4, 6, 8, \dots\}$ closed under the binary operation of addition?

☐ yes, ☐ no   Reason: Let $2m$ and $2n$ be arbitrary elements in E.

Then since ...

How about under multiplication?

Task 2 Is $A = \{0, 1, 4, 9, 16, \dots\}$ closed under addition?

Under subtraction?

Under multiplication?

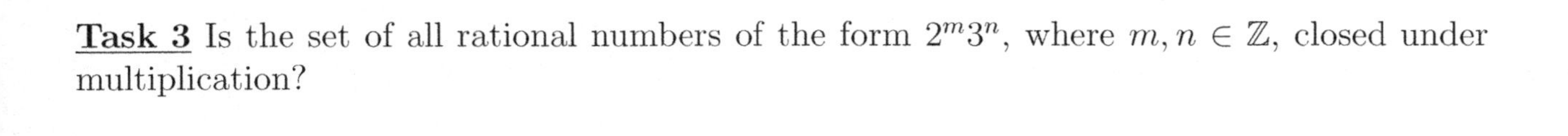

Task 3 Is the set of all rational numbers of the form $2^m 3^n$, where $m, n \in \mathbb{Z}$, closed under multiplication?

Task 4 Is the set of all positive rational numbers closed under addition? Multiplication?

Task 5 Are the complex numbers of the form $m + ni$ where m and n are integers closed under multiplication?

Task 6 Is the set $\{m + n\sqrt{2} : m, n \in \mathbb{Z}\}$ closed under multiplication?

Task 7 Are the irrationals closed under multiplication? Under subtraction?

WORKSHEET ON CLOSURE II

Let $\mathbb{Z} = \{\ldots, -2, -1, 0, 1, 2, \ldots\}$
QUESTION: Which of the sets $3\mathbb{Z}$, $1 + 3\mathbb{Z}$, $2 + 3\mathbb{Z}$ are closed under subtraction?

Task 1 List the elements of $3\mathbb{Z}$; choose two and subtract them.

Task 2 What does it mean to say $3\mathbb{Z}$ is closed under subtraction?

Task 3 Is $3\mathbb{Z}$ closed under subtraction? If yes, prove it.

Task 4 Is $1 + 3\mathbb{Z}$ closed under subtraction?

Task 5 Is $2 + 3\mathbb{Z}$ closed under subtraction?

Task 6 Why must a set of integers contain 0 to be closed under subtraction?

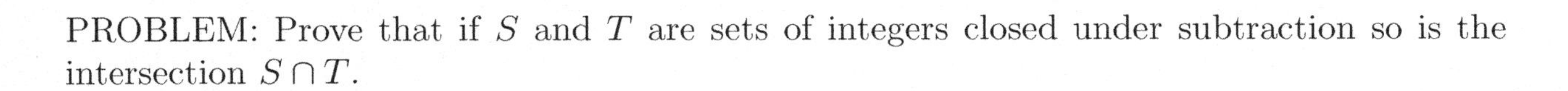

WORKSHEET ON CLOSURE III

PROBLEM: Prove that if S and T are sets of integers closed under subtraction so is the intersection $S \cap T$.

Task 1 Say in your own words what it means to say S is closed under subtraction.

Task 2 What do you have to show in order to check that $S \cap T$ is closed under subtraction?

Task 3 Draw a Venn diagram as an aid, and resolve the problem.

Task 4 If S and T are sets of integers closed under subtraction is the union $S \cup T$ also closed under subtraction? If yes, prove it, if no give a counterexample.

WORKSHEET ON RESIDUE CLASSES

Congruence Modulo m is an EQUIVALENCE RELATION on $\mathbb{Z}$, the set of all integers.

R – REFLEXIVE: $a \equiv a(m)$
S – SYMMETRIC: If $a \equiv b(m)$ then $b \equiv a(m)$
T – TRANSITIVE: $a \equiv b(m)$ and $b \equiv c(m)$ then $a \equiv c(m)$

The relation *congruence* partitions $\mathbb{Z}$ into disjoint EQUIVALENCE CLASSES or RESIDUE CLASSES.

When $m = 2$, $\mathbb{Z}$ is partitioned into the classes $2\mathbb{Z}$ and $1 + 2\mathbb{Z}$.

$$2\mathbb{Z} = \{\ldots, -4, -2, 0, 2, 4, \ldots\}$$
$$1 + 2\mathbb{Z} = \{\ldots, -3, -1, 1, 3, 5, 7, \ldots\}$$

Task 1 Explain why the classes $2\mathbb{Z}$ and $1 + 2\mathbb{Z}$ are disjoint.

Task 2 What the residue classes when $m = 3$? Are they disjoint? Why?

Task 3 When $m = 4$? Explain.

WORKSHEET ON MODULAR ARITHMETIC

$a \equiv b(\text{mod } m)$ means a and b have the same remainder when divided by m; or that $a - b$ is divisible by m, or $a - b = mk$. An example: $17 \equiv 9(\text{mod } 4)$ since 4 divides $17 - 9$.

Task 1 Complete the missing four rows:

	0	1	2	3	4	5	6	7	8	9	10	11	12
Mod 2													
Mod 3	0	1	2	0	1	2	0	1	2	0	1	2	0
Mod 4													
Mod 5													
Mod 6													

Task 2 Tables of addition Mod 5 and Mod 6 would look like:

$\oplus$	0	1	2	3	4
0					
1	1	2	3	4	0
2					
3					
4					

$\oplus$	0	1	2	3	4	5
0						
1						
2						
3						
4						
5						

Let $\mathbb{Z}_6 = \{0, 1, 2, 3, 4, 5\}$ be the six elements you used to make the 6 by 6 table in Task 2. If you examine that addition table, you can see that each of the following subsets are closed under the binary operation $\oplus$. The relationship among these subsets is shown in the diagram.

$$A = \{0\}$$
$$B = \{0, 3\}$$
$$C = \{0, 2, 4\}$$
$$D = \{0, 1, 2, 3, 4, 5\}$$

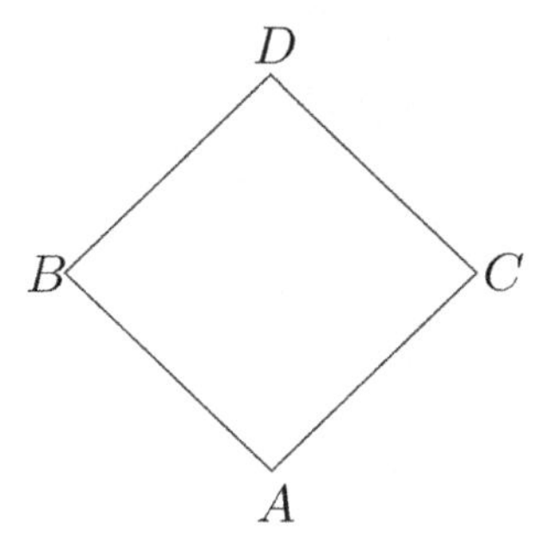

<u>**Task 3**</u> Make the $\oplus$ table for $\mathbb{Z}_8 = \{0, 1, 2, 3, 4, 5, 6, 7\}$, list all the subsets closed under $\oplus$, and make a diagram as above.

$\oplus$	0	1	2	3	4	5	6	7
0								
1								
2								
3								
4								
5								
6								
7								

$A = \{0\}$

$B =$

$C =$

$D =$

<u>**Task 4**</u> Without making the addition table, can you give all the closed subsets of $\mathbb{Z}_{12}$?

WORKSHEET ON CANCELLATION

Cancellation Theorem: If either $ab = ac$ or $ba = ca$ in a group G, then $b = c$.

Task 1 Let's try to prove right cancellation.

The hypothesis for right cancellation is:

If the element a is in G, ____________________ is also in G.

Now show how to use this latter element on $ba = ca$ and conclude that $b = c$.

– CONNECTIONS –

Task 2 Let A, B, C be sets in a universe S. If $A \cup B = A \cup C$ is it necessarily true that $B = C$?

Task 3 Does $A \cap B = A \cap C$ imply $B = C$?

Task 4 For 2 by 2 matrices A, B, C does $AB = AC$ imply $B = C$?

Task 5 For real numbers x, y, z does $x + y = x + z$ imply $y = z$?

PERMUTATIONS

Each permutation on $X_4 = \{1,2,3,4\}$ is a 1–1, onto function f. For example, the permutation $1 \to 2$, $2 \to 4$, $3 \to 3$, $4 \to 1$ has the function table

x	1	2	3	4
$f(x)$	2	4	3	1

and can be expressed in cycle form as (124). With "multiplication" being composition of functions the product (124)(23) is (1324), operating left to right. The cycle form (124) means $1 \to 2 \to 4 \to 1$ with 3 fixed. The product (124)(23) can be written as two-rowed arrays as:

$$\begin{pmatrix} 1 & 2 & 3 & 4 \\ 2 & 4 & 3 & 1 \end{pmatrix}\begin{pmatrix} 1 & 2 & 3 & 4 \\ 1 & 3 & 2 & 4 \end{pmatrix} = \begin{pmatrix} 1 & 2 & 3 & 4 \\ 3 & 4 & 2 & 1 \end{pmatrix} = (1324)$$

We list cycles in standard form as follows:

1. Smallest number first
2. Omission of a number m means $m \to m$ is fixed

Task 1

a=(1342)

$a^2 =$

$a^3 =$

$a^4 =$

a=(24675)

$a^2 =$

$a^3 =$

$a^4 =$

$a^5 =$

Task 2 If β =(26), $\beta^{-1} =$

What is the inverse of any transposition (ab)?____________________

Task 3 Let α=(132)(4675). What is the smallest positive integer s so that

$\alpha^s = (1)$? s=____________________________.

Repeat with β=(12)(3465): s=____________________________.

Give the order of each element by filling in the chart:

α	(13)	(132)	(12)(34)	(1432)	(132)(23)	(13)(12)
order α						

What is the order of $\beta = (13)(257)(4689)$?

What is the order of $\gamma = (13)(234)$?

WORKSHEET ON SUBGROUPS

Let $H = \{\alpha : \alpha = a + bi, |\alpha| \leq 1\}$. Is H a subgroup of the multiplicative group of non-zero complex numbers?

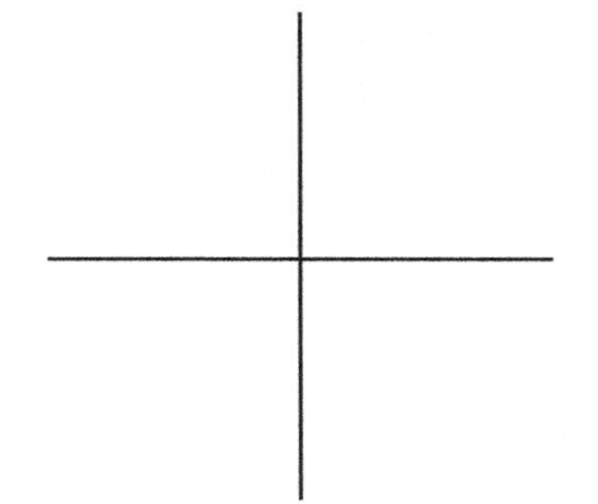

<u>**Task 1**</u> Draw a picture showing all α with $|\alpha| \leq 1$.
Recall $|a + bi| = \sqrt{a^2 + b^2}$.

<u>**Task 2**</u> To show that H is a subgroup we need to show, for one thing, that if $\alpha \in H$ so is α^{-1}. What is the inverse of $a + bi$? Try this on $\alpha = \frac{1}{2} + \frac{1}{2}i$. What is α^{-1}?

Is $\alpha \in H$? You need to compute $\sqrt{\frac{1}{4} + \frac{1}{4}}$.

Draw α and α^{-1} in your picture as vectors. How are their angles related? How do you multiply two complex numbers to show $\alpha\alpha^{-1} = 1$?

<u>**Task 3**</u> Give an α <u>not</u> in H.

<u>**Task 4**</u> Do you now need to check closure?

WORKSHEET ON ORDER

The order of an element a of a group G is the order of the cyclic subgroup $[a]$ generated by a in G. Equivalently, it is the smallest positive integer m so that $a^m = e$.

Task 1 Let G be a cyclic group of order 18 generated by a. Then

$$G = \{e, a, a^2, a^3, a^4, a^5, a^6, a^7, a^8, a^9, a^{10}, a^{11}, a^{12}, a^{13}, a^{14}, a^{15}, a^{16}, a^{17}\}.$$

Give the order of each element of G by filling out the chart:

x	e	a	a^2	a^3	a^4	a^5	a^6	a^7	a^8
order x									

x	a^9	a^{10}	a^{11}	a^{12}	a^{13}	a^{14}	a^{15}	a^{16}	a^{17}
order x									

Task 2 Repeat Task 1 with G being a cyclic group of order 24.

WORKSHEET ON GROUP TABLES

Give as many reasons as you can why each of these tables cannot be group operation tables. You can state a group axiom that fails, or appeal to some of our theorems and results.

	a	b	c	d	e
a	c	e	a	b	d
b	d	c	b	e	a
c	a	b	c	d	e
d	e	a	d	c	b
e	b	d	e	a	c

	a	b	c	d	e
a	c	e	a	b	d
b	d	a	b	e	c
c	a	b	c	d	e
d	e	c	d	a	b
e	b	d	e	c	a

	a	b	c	d	e
a	e	d	b	c	a
b	c	e	d	a	b
c	d	a	e	b	c
d	b	c	a	e	d
e	a	b	c	d	e

WORKSHEET ON COMPLEX NUMBERS

Task 1 $i^2 =$

$(1+i)(2-3i) =$

Task 2 Locate each of the following in the cartesian plane.

(a) 1, $(-1+i\sqrt{3})/2$, $(-1-i\sqrt{3})/2$

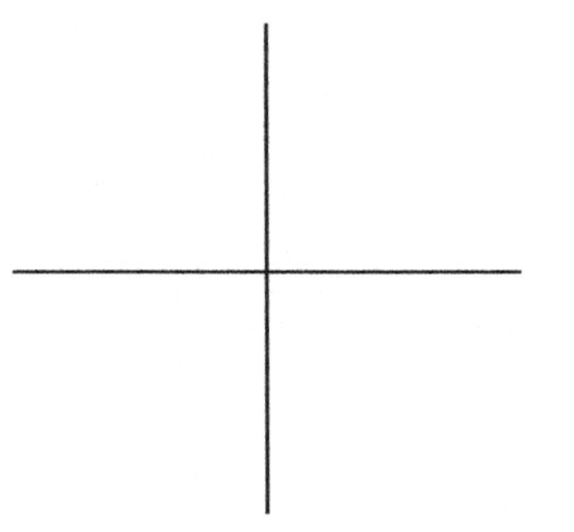

(b) 1, -1, i, $-i$

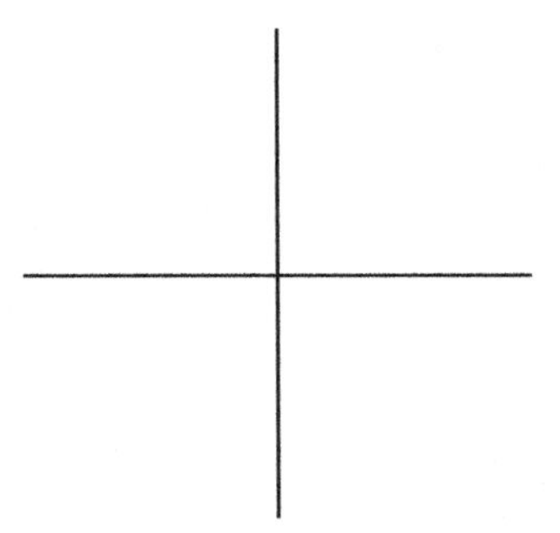

(c) Connect each of the points in (a) and describe the properties of the figure. Repeat with (b).

(d) What are the roots of $x^3 - 1 = 0$, $x^4 - 1 = 0$, and how is this question related to the above parts?

<u>**Task 3**</u> If r and s are roots of $x^2 - 7x + 43 = 0$, what are $r + s$ and rs?

<u>**Task 4**</u> Use $(x - a)(x - b) = x^2 - (a + b)x + ab$ to resolve Task 3.

Give a similar expression for $(x - a)(x - b)(x - c)$.

What is the sum of the roots of $x^3 - 3x^2 + 2x - 14 = 0$? The product of the roots?

Let 1, r, s be roots of $x^3 - 1 = 0$.

The product of the roots is $1rs =$______________, so that $rs =$______________.

Also, $r^3 =$______________, $s^3 =$______________.

Explain why $r^2 = -r - 1$ and $r = \frac{1}{s}$ and $r^3 = \frac{r^2}{s}$.

Why is $r^2 = s$?

MULTIPLICATION TABLE OF ROOTS

Task 1 Use $x^n - 1 = (x-1)(x^{n-1} + x^{n-2} + \cdots + x + 1)$ to complete the chart:

	FACTORS	ROOTS
$x^2 - 1 = 0$	$(x-1)(x+1)$	1,−1
$x^3 - 1 = 0$		
$x^4 - 1 = 0$		

For convenience, you might want to label the roots of $x^3 - 1 = 0$ as 1, β, γ.

Task 2 Make the multiplication table for each set of roots:

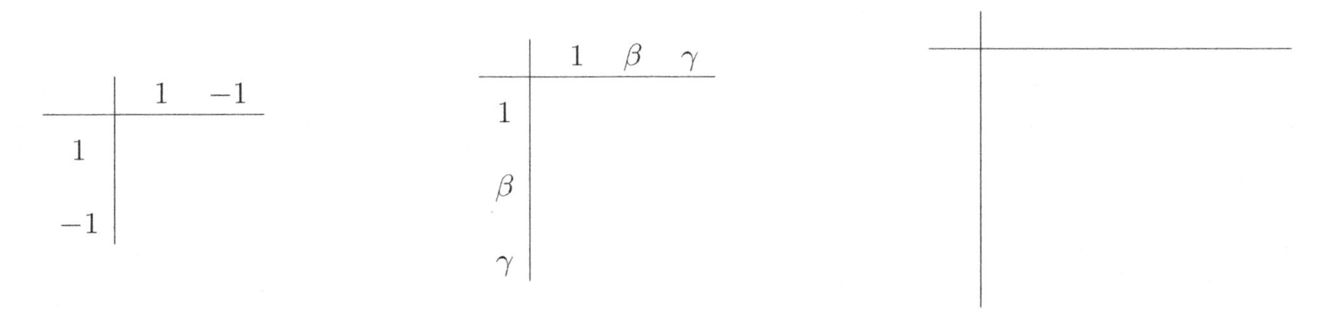

Task 3 Plot separately the set of roots for $x^2 - 1 = 0$, $x^3 - 1 = 0$, and $x^4 - 1 = 0$.

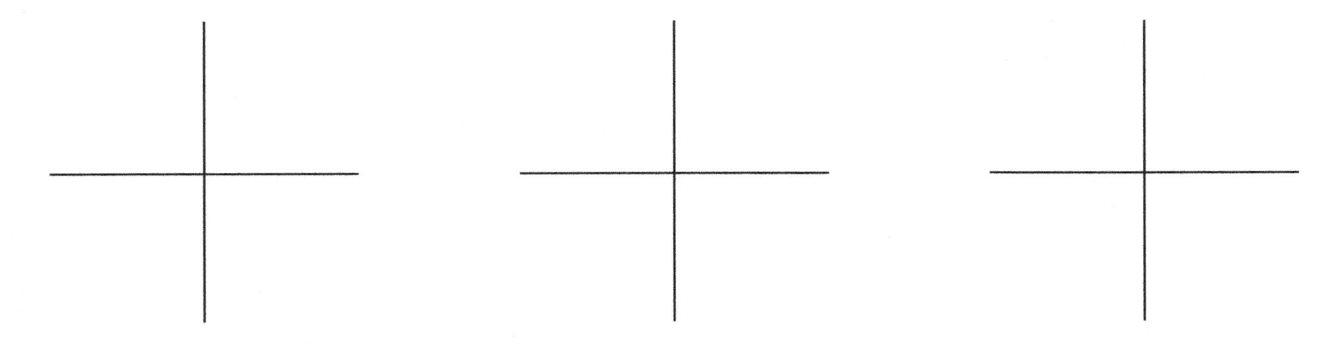

Connect the points and describe the geometrical figure produced.

Task 4 Conjecture what happens for $x^5 - 1 = 0$, $x^6 - 1 = 0$.

GROUP TABLE FOR THE SIXTH ROOTS OF UNITY

The goal here is to find the six roots of $x^6 - 1 = 0$, plot them, and make their group table.

Task 1 Factor $x^6 - 1 = (x^3 - 1)(\qquad) = (\qquad)(\qquad)(\qquad)(\qquad)$

The six roots are:

Task 2 Plot these six complex numbers.

Connect them, forming a ____________________

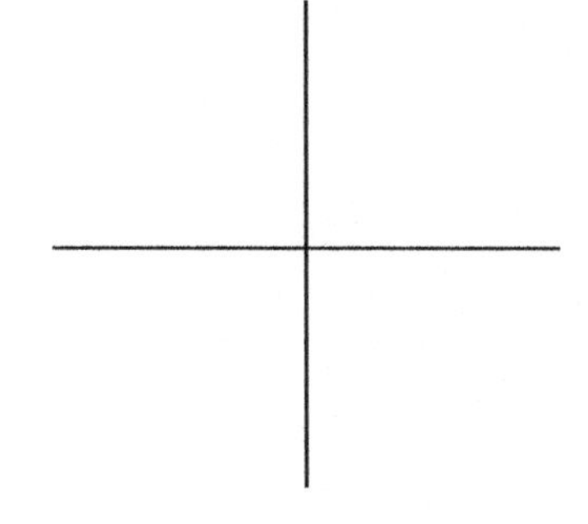

Task 3 Label the six roots 1, r, s, -1, $-r$, $-s$ where 1, r, s are the roots of $x^3 - 1 = 0$. Show why the last three are negatives of the first three. Show why $rs = 1$, $r^2 = s$, $s^2 = r$.

Task 4 Make the group table

	1	r	s	-1	$-r$	$-s$
1						
r						
s						
-1						
$-r$						
$-s$						

MULTIPLICATION TABLE FOR THE ROOTS OF $x^8 - 1 = 0$

FACTOR: $x^8 - 1 = (x^4 - 1)(x^4 + 1)$. The roots of $x^4 - 1 = 0$ are ____________________

$$x^4 + 1 = x^4 + 2x^2 + 1 - 2x^2 = (x^2 + 1)^2 - (\sqrt{2}\,x)^2 = (x^2 - \sqrt{2}\,x + 1)(x^2 + \sqrt{2}\,x + 1)$$

The other roots are:

$$r = \frac{1}{\sqrt{2}} + \frac{i}{\sqrt{2}} \qquad -r = \frac{-1}{\sqrt{2}} - \frac{i}{\sqrt{2}}$$

$$s = \frac{-1}{\sqrt{2}} + \frac{i}{\sqrt{2}} \qquad -s = \frac{1}{\sqrt{2}} - \frac{i}{\sqrt{2}}$$

These eight roots are spread evenly around a unit circle 45° apart.

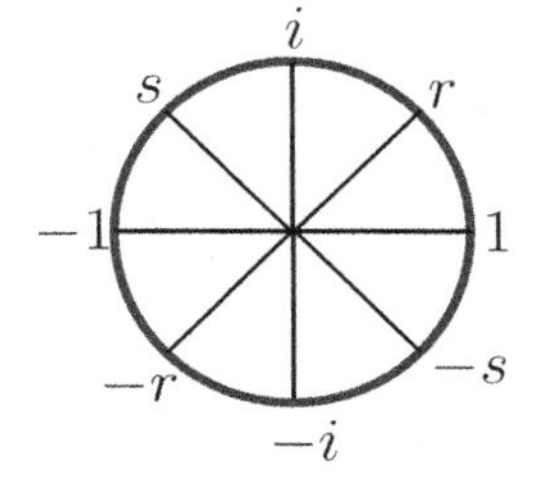

Task 1 Use the fact that when you multiply complex numbers you add their arguments to express each of the following in terms of 1, -1, i, $-i$, r, s, $-r$, $-s$.

$r^2 =$ $\qquad$ $ir =$ $\qquad$ $s^2 =$

$rs =$ $\qquad$ $is =$

Task 2 Complete the multiplication table

	1	-1	i	$-i$	r	s	$-r$	$-s$
1								
-1								
i								
$-i$								
r								
s								
$-r$								
$-s$								

COMPOSITION OF FUNCTIONS

Let $f_1(x)=x$ $\quad f_4(x)=\dfrac{1}{x}$

$f_2(x)=\dfrac{1}{1-x}$ $\quad f_5(x)=1-x$

$f_3(x)=\dfrac{x-1}{x}$ $\quad f_6(x)=\dfrac{x}{x-1}$

The following "multiplication" table can be formed using composition of functions as the operation

EXAMPLE: $f_2 \circ f_6 = f_5$ since $f_2\left(\dfrac{x}{x-1}\right) = \dfrac{1}{1-\frac{x}{x-1}} = 1-x = f_5(x)$

$\circ$	x	$\dfrac{1}{1-x}$	$\dfrac{x-1}{x}$	$\dfrac{1}{x}$	$1-x$	$\dfrac{x}{x-1}$
$f_1 = x$	x	$\dfrac{1}{1-x}$	$\dfrac{x-1}{x}$	$\dfrac{1}{x}$	$1-x$	$\dfrac{x}{x-1}$
$f_2 = \dfrac{1}{1-x}$	$\dfrac{1}{1-x}$	$\dfrac{x-1}{x}$	x	$\dfrac{x}{x-1}$	$\dfrac{1}{x}$	$1-x$
$f_3 = \dfrac{x-1}{x}$	$\dfrac{x-1}{x}$					
$f_4 = \dfrac{1}{x}$	$\dfrac{1}{x}$					
$f_5 = 1-x$	$1-x$					
$f_6 = \dfrac{x}{x-1}$	$\dfrac{x}{x-1}$					

COMPOSITION OF FUNCTIONS (CONT)

Rewrite the table using $f_1, f_2, f_3, f_4, f_5, f_6$

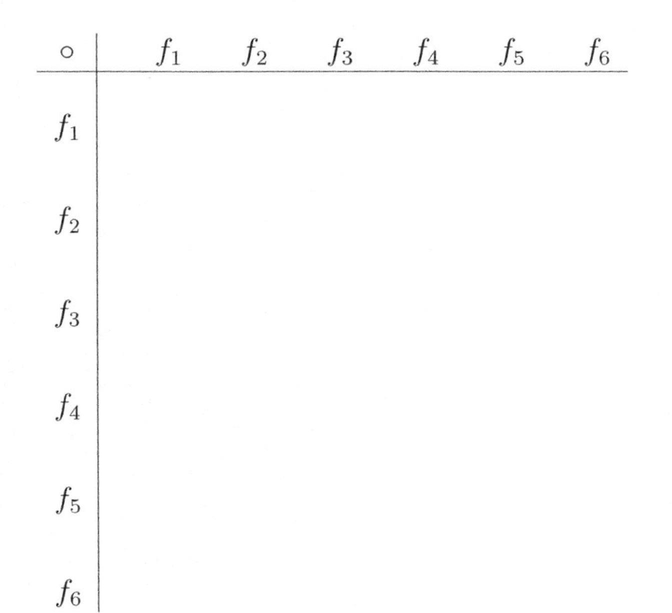

$\circ$	f_1	f_2	f_3	f_4	f_5	f_6
f_1						
f_2						
f_3						
f_4						
f_5						
f_6						

List properties of this table.

Complete the table showing inverses

f	f_1	f_2	f_3	f_4	f_5	f_6
f^{-1}						

List all the subsets of $\{f_1, f_2, f_3, f_4, f_5, f_6\}$ that are closed under $\circ$.

EULER ϕ-FUNCTION

$\phi(n)$ is the number of positive integers less than n that are relatively prime to n. Here is a partial table:

n	1	2	3	4	5	6	7	8	9	10	11	12	13	14	15	16
$\phi(n)$	1	1	2	2	4	2	6	4	6	4	10					

Task 1 Complete the table. Any conjectures?

Task 2 Conjecture and prove a formula for $\phi(p)$, p a prime.

Task 3 Prove a formula for $\phi(p^2)$.

Task 4 Compute $\phi(7^3)$ by listing the integers.

Task 5 Prove a formula for $\phi(p^n)$.

Task 6 Prove that $\phi(11^n)$ is a multiple of 10, for all n.

Task 7 Show that $\phi(16) \cdot \phi(9) = \phi(16 \cdot 9)$

<u>**Task 8**</u> Find all x such that $\phi(x) = n$ where:

(a) n =1 (b) n=2 (c) n=4

<u>**Task 9**</u> The notation $(a, b) = 1$ means that a and b are relatively prime.
Prove that if $(a, m) = 1$, then $(m - a, m) = 1$.

<u>**Task 10**</u> Prove that $\phi(n)$ is even for $n \geq 3$.

<u>**Task 11**</u> It can be proved that if $(m, n) = 1$, then $\phi(mn) = \phi(m)\phi(n)$.
Use this to compute $\phi(72)$; also compute $\phi(120)$.

<u>**Task 12**</u> Prove that if $n = p_1^{e_1} p_2^{e_2} p_3^{e_3}$, then $\phi(n) = n\left(1 - \frac{1}{p_1}\right)\left(1 - \frac{1}{p_2}\right)\left(1 - \frac{1}{p_3}\right)$.

WORKSHEET ON INVERTIBLES

Let $\mathbb{Z}_m = \{\bar{0}, \bar{1}, \bar{2}, \bar{3}, \ldots, \overline{m-1)}\}$ and V_m be the set of invertibles of $\mathbb{Z}_m$ consisting of those elements of $\mathbb{Z}_m$ that have *multiplicative* inverses. For each $\mathbb{Z}_m$ make a table of inverses of those elements that have multiplicative inverses and list V_m. Here the "bar" indicates an equivalence class. $\bar{2}$ indicates the set of all integers in $\mathbb{Z}$ whose remainder is 2 upon division by m. Once understood, the "bar" is omitted.

SAMPLE:

$\mathbb{Z}_3 = \{0, 1, 2\}$

x	0	1	2
x^{-1}		1	2

$V_3 = \{1, 2\}$

$\mathbb{Z}_4 = \{0, 1, 2, 3\}$

x	0	1	2	3
x^{-1}				

$V_4 = \{ \qquad \}$

$\mathbb{Z}_5 = \{0, 1, 2, 3, 4\}$

x	0	1	2	3	4
x^{-1}					

$V_5 = \{ \qquad \}$

$\mathbb{Z}_6 = \{0, 1, 2, 3, 4, 5\}$

x	0	1	2	3	4	5
x^{-1}						

$V_6 = \{ \qquad \}$

$\mathbb{Z}_7 = \{0, 1, 2, 3, 4, 5, 6\}$

x	0	1	2	3	4	5	6
x^{-1}							

$V_7 = \{ \qquad \}$

Can you conjecture which elements of $\mathbb{Z}_{30}$ are invertibles?

How many invertibles does $\mathbb{Z}_p$ have where p is a prime?

How is the Euler ϕ-function related to these questions?

PRESERVATION OF OPERATION

In the following chart you are asked to verify whether certain familiar functions satisfy

$f(a \circ b) = f(a) \square f(b)$. The operations $\circ$ and $\square$ can be addition or multiplication or a mixture.

FUNCTION	YES OR NO	REASON
$f(x) = x^3$	yes	$f(xy) = (xy)^3 = x^3y^3 = f(x)f(y)$
$f(x) = x^4$		
$f(x) = e^x$		
$f(x) = \frac{3}{2}x$		
$f(x) = 2x + 1$		
$f(x) = \ln x$		
$f(x) = \|x\|$		
$f(x) = \sqrt{x}$		
$f(x) = 2x^3$		
$f(x) = \det x$		
$\theta(f) = f'$ (the derivative)		

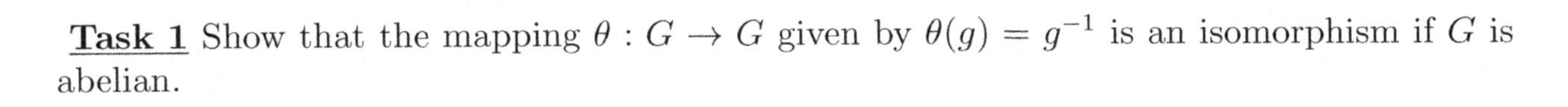

A SPECIAL ISOMORPHISM

Task 1 Show that the mapping $\theta : G \to G$ given by $\theta(g) = g^{-1}$ is an isomorphism if G is abelian.

STEP 1 θ is 1-to-1. To show this we need to show that if $\theta(g_1) = \theta(g_2)$ then

____________________________________. Since $\theta(g_1) = \theta(g_2)$ we get

____________________________________. Now by taking inverses, we obtain

____________________________________.

STEP 2 Show that θ is onto.

STEP 3 Show that θ preserves the operation

Task 2 Use $\theta(g) = g^{-1}$ to tabulate an isomorphism from S_3, the group of symmetries of an equilateral triangle, to itself.

g	
$\theta(g)$	

Task 3 Show that the above result is false if G is not abelian.

MATCHING GROUPS

There are two nonisomorphic groups of order 4, the cyclic group and the Klein 4-group whose elements x satisfy $x^2 = e$.

For each of the following groups, label A if it is isomorphic to the cyclic group and B if it is isomorphic to the Klein group.

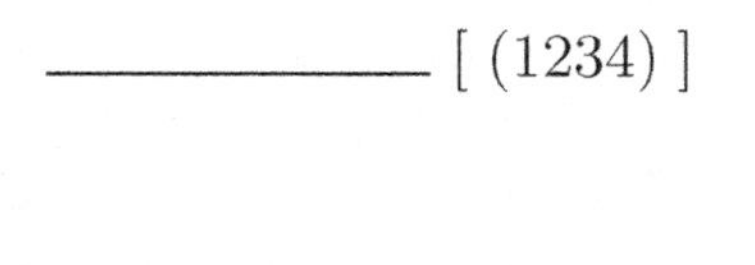

__________ $[\,(1234)\,]$

A. Cyclic
B. Klein

__________ $[-i]$

__________ $\{e, a, a^2, a^3\}$

__________ $[i]$

__________ Rectangle Group

__________ $\{1, -1, i, -i\}$

__________ $\{0, 1, 2, 3\}$ under addition mod 4

__________ $\{(1), (12), (34), (12)(34)\}$

WORKSHEET ON AN 8×8 GROUP TABLE – $\mathbb{Z}_8$

Task 1 Fill in the following table where each of the 64 entries is found by addition modulo 8; i.e. add the two numbers, divide by 8, and record the remainder.

$\oplus$	0	1	2	3	4	5	6	7
0								
1								
2								
3								
4								
5								
6								
7								

MAKE A TABLE OF INVERSES

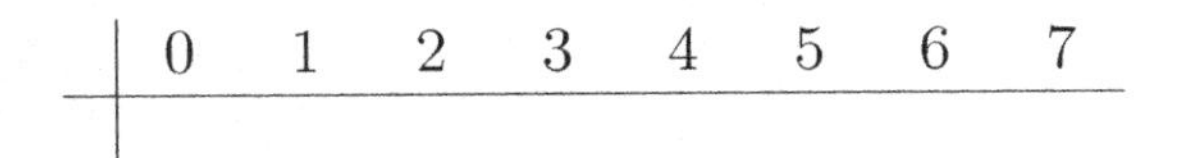

	0	1	2	3	4	5	6	7

DRAW THE LATTICE OF SUBGROUPS.

LABEL AND LIST ALL THE SUBGROUPS.

WORKSHEET ON AN 8×8 GROUP TABLE – $\mathbb{Z}_2 \times \mathbb{Z}_4$

Task 1 Fill in the following table using bitwise addition mod 2 in the left-most slot and bitwise addition mod 4 in the right-most slot.

$\oplus$	00	01	02	03	10	11	12	13
00								
01								
02								
03								
10								
11								
12								
13								

MAKE A TABLE OF INVERSES

DRAW THE LATTICE OF SUBGROUPS.

LABEL AND LIST ALL THE SUBGROUPS.

WORKSHEET ON AN 8×8 GROUP TABLE – $\mathbb{Z}_2 \times \mathbb{Z}_2 \times \mathbb{Z}_2$

Task 1 Fill in the following table using bitwise addition mod 2.

$\oplus$	000	001	010	011	100	101	110	111
000								
001								
010								
011								
100								
101								
110								
111								

MAKE A TABLE OF INVERSES

DRAW THE LATTICE OF SUBGROUPS.

LABEL AND LIST ALL THE SUBGROUPS.

WORKSHEET ON AN 8×8 GROUP TABLE – THE QUATERNIONS

Task 1 Fill in the following table using the operations:

$$i^2 = j^2 = k^2 = -1,\ ij = -ji = k,\ jk = -kj = i,\ ki = -ik = j$$

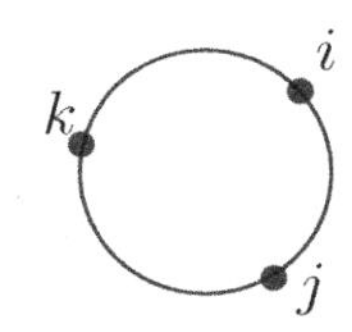

$\otimes$	1	-1	i	$-i$	j	$-j$	k	$-k$
1								
-1								
i								
$-i$								
j								
$-j$								
k								
$-k$								

MAKE A TABLE OF INVERSES

DRAW THE LATTICE OF SUBGROUPS.

LABEL AND LIST ALL THE SUBGROUPS.

WORKSHEET ON AN 8×8 GROUP TABLE – THE OCTIC GROUP

Task 1 Fill in the following table using $fr = r^3f$. These eight elements are the eight symmetries of a square.

$\square$	1	r	r^2	r^3	f	rf	r^2f	r^3f
1								
r								
r^2								
r^3								
f								
rf								
r^2f								
r^3f								

MAKE A TABLE OF INVERSES

DRAW THE LATTICE OF SUBGROUPS.

LABEL AND LIST ALL THE SUBGROUPS.

FIVE NONISOMORPHIC GROUPS OF ORDER 8

Listed next are the elements of these five groups along with their names. You are asked to show why certain pairs are not isomorphic.

CYCLIC	$\{e, a, a^2, a^3, a^4, a^5, a^6, a^7\}$
QUATERNIONS	$\{1, -1, i, -i, j, -j, k, -k\}$
OCTIC	$\{(1), \rho, \rho^2, \rho^3, \phi, \rho\phi, \rho^2\phi, \rho^3\phi\}$
BIT STRINGS or $\mathbb{Z}_2 \times \mathbb{Z}_2 \times \mathbb{Z}_2$	$\{000, 001, 010, 011, 100, 101, 110, 111\}$
$\mathbb{Z}_2 \times \mathbb{Z}_4$	$\{00, 01, 02, 03, 10, 11, 12, 13\}$

Task 1 Give two reasons why the QUATERNIONS are not isomorphic to the OCTIC group.

Task 2 Why is $\mathbb{Z}_2 \times \mathbb{Z}_4$ not isomorphic to $\mathbb{Z}_2 \times \mathbb{Z}_2 \times \mathbb{Z}_2$?

Task 3 Why is the CYCLIC group not isomorphic to any of the other four?

WORKSHEET ON GROUP TABLES, ISOMORPHISMS

Complete the following table using $\circ$ to mean composition of functions. For example.

$$f_2 \circ f_3 = f_2(f_3(x)) = f_2(1-x) = \frac{1}{1-x} = f_4$$

The six functions are:

$$f_1(x) = x \quad f_2(x) = \frac{1}{x} \quad f_3(x) = 1-x \quad f_4(x) = \frac{1}{1-x} \quad f_5(x) = \frac{x-1}{x} \quad f_6 = \frac{x}{x-1}$$

$\circ$	f_1	f_2	f_3	f_4	f_5	f_6
f_1						
f_2						
f_3						
f_4						
f_5						
f_6						

Display an isomorphism between this group and either S_3 or $\mathbb{Z}_6$.

FUNDAMENTAL THEOREM OF FINITE ABELIAN GROUPS

THEOREM: Every finite abelian group can be written as a product of cyclic groups of prime power order.

EXAMPLES: The Klein 4–Group is $\mathbb{Z}_2 \times \mathbb{Z}_2$; the cyclic group of order 4 is $\mathbb{Z}_4$. The abelian group of order 6 is $\mathbb{Z}_6$ which is isomorphic to the direct product $\mathbb{Z}_2 \times \mathbb{Z}_3$.

Let $G = \{1, 8, 12, 14, 18, 21, 27, 31, 34, 38, 44, 47, 51, 53, 57, 64\}$ be a group of order 16 under multiplication mod 65. G is isomorphic to one of:

$\mathbb{Z}_{16}$
$\mathbb{Z}_2 \times \mathbb{Z}_8$
$\mathbb{Z}_4 \times \mathbb{Z}_4$
$\mathbb{Z}_2 \times \mathbb{Z}_2 \times \mathbb{Z}_4$
$\mathbb{Z}_2 \times \mathbb{Z}_2 \times \mathbb{Z}_2 \times \mathbb{Z}_2$

BUT WHICH ONE?

LOOK AT ORDERS!

x	1	8	12	14	18	21	27	31	34	38	44	47	51	53	57	64
order x	1	4	4	2		4	4			4	4	4		4	4	

Task 1 Why is G not $\mathbb{Z}_{16}$?

Task 2 Why is G not $\mathbb{Z}_2 \times \mathbb{Z}_8$?

Task 3 What are the orders of elements in $\mathbb{Z}_2 \times \mathbb{Z}_2 \times \mathbb{Z}_2 \times \mathbb{Z}_2$?

Task 4 Which one must G be?

Task 5 $G = \{1, 9, 16, 22, 29, 53, 74, 79, 81\}$ is a group of order 9 under multiplication modulo 91. Is G isomorphic to $\mathbb{Z}_9$ or $\mathbb{Z}_3 \times \mathbb{Z}_3$? Why are these the only two possibilities?

Make the table of orders and inverses.

x	1	9	16	22	29	53	74	79	81
Order x									

x	1	9	16	22	29	53	74	79	81
x^{-1}									

Task 6 Identify all abelian groups (up to isomorphism) of order 360 by doing the following:

A. Express 360 as a product of prime numbers

B. List the six direct product possibilities

WHICH DIRECT PRODUCT?

$V_{45} = \{1, 2, 4, 7, 8, 11, 13, 14, 16, 17, 19, 22, 23, 26, 28, 29, 31, 32, 34, 37, 38, 41, 43, 44\}$ is a multiplicative group, using mod 45, of order 24. According to the fundamental theorem of finite abelian groups, V_{45} is isomorphic to a direct product of cyclic groups, each having prime power order. The possibilities are:

$$\mathbb{Z}_3 \times \mathbb{Z}_8 \qquad \mathbb{Z}_2 \times \mathbb{Z}_3 \times \mathbb{Z}_4 \qquad \mathbb{Z}_2 \times \mathbb{Z}_2 \times \mathbb{Z}_2 \times \mathbb{Z}_3$$

You do not include $\mathbb{Z}_{24}$, $\mathbb{Z}_6 \times \mathbb{Z}_4$ or $\mathbb{Z}_2 \times \mathbb{Z}_{12}$ since the orders are not prime power. Also notice that the cyclic group $\mathbb{Z}_{24}$ is, in fact, $\mathbb{Z}_3 \times \mathbb{Z}_8$ which has (1, 1) as a generator.

One way of determining which of these three is isomorphic to V_{45} is by computing the orders of each element in V_{45} and comparing with orders of elements in the direct products. By hand, this is not an easy feat.

USING THE TI-92

Clear home screen – F1, 8
Clear input line – CLEAR
Type in

Define $f(n) = \text{mod}(22^n, 45)$

to determine the order of 22, eg., and enter.
Go to APPS and 6: Data/matrix editor, current and enter
If necessary clear columns with F6, 5.
Highlight C1 and enter. Generate the sequence 1, 2, ..., 24 with

C1 = seq(n, n, 1, 24) and enter.

Highlight C2 and generate $f(n)$ with

C2 = seq($f(n)$, n, 1, 24) and enter.

Column C2 will give 22^n reduced modulo 45 for n ranging from 1 to 24. You should find that $22^{12} \equiv 1$.
Now, if you return to the home screen [2nd, QUIT] and just change the 22 to 2 we can quickly see, returning to APPS, 6, in column 2 that the order of 2 is 12. Repeat this and complete the following table of orders.

x	1	2	4	7	8	11	13	14	16	17	19	22
order x	1	12	6	12								

x	23	26	28	29	31	32	34	37	38	41	43	44
order x											12	2

RINGS THAT ARE NOT INTEGRAL DOMAINS

A commutative ring D with unity 1, having no zero-divisors is called an integral domain.

Task 1 Explain why $\mathbb{Z}_{10}$ is not an integral domain.

Task 2 Why is $\mathbb{Z}_{12}$ not?

Task 3 For which m is $\mathbb{Z}_m$ not an integral domain?

Task 4 Is $\mathbb{Z}_2 \times \mathbb{Z}_2$ an integral domain?

Task 5 Let A and B be integral domains. Is $A \times B$ an integral domain?

Task 6 Is the set of all 2×2 matrices with real entries with the usual addition an multiplication of matrices an integral domain?

WORKSHEET ON POLYNOMIALS IN $\mathbb{Z}_n[x]$

Task 1 Tabulate each of $(x+\bar{2})(x+\bar{5})$ and $x(x+\bar{7})$ in $\mathbb{Z}_{10}$ and thus show that they are equal.

x	0	1	2	3	4	5	6	7	8	9
$(x+\bar{2})(x+\bar{5})$			8							
$x(x+\bar{7})$			8							

Task 2 Show that $(x+\bar{3})(x+\bar{5}) = x(x+\bar{8})$ in $\mathbb{Z}_{15}[x]$.

Task 3 Show that $(x+\bar{6})^2 = x^2$ in $\mathbb{Z}_{12}[x]$.

Task 4 Our experience leads us to expect that deg $\alpha\beta$ = deg α + deg β

for polynomials α and β.

But ... Show that $(\bar{2}x+\bar{1})(\bar{3}x+\bar{5}) = x+\bar{5}$ in $\mathbb{Z}_6[x]$.

Task 5 Show that in $\mathbb{Z}_6[x]$

$$(\bar{2}x + \bar{5})(\bar{3}x + \bar{2}) = (x + \bar{4})$$

$$(\bar{3}x + \bar{3})(\bar{4}x^2 + \bar{2}) = \bar{0}$$

Task 6 In $\mathbb{Z}_5[x]$, $(\bar{2}x + \bar{1})(\bar{4}x + \bar{3}) = (x + \bar{3})(\bar{3}x + \bar{1})$

Make these linear factors monic (leading coefficient is $\bar{1}$) by factoring out $\bar{2}, \bar{4}$, and $\bar{3}$. For example $(\bar{2}x + \bar{1}) = \bar{2}(x + \bar{3})$. Then both sides become

________________________.

WORKSHEET ON ANOTHER RING

Define "addition" and "multiplication" as follows:

$$a \oplus b = a + b - 1$$

$$a \otimes b = a + b - ab$$

show that $(\mathbb{Z}, \oplus, \otimes)$ is a ring by doing the following:

Task 1 Determine the "0", the additive identity. Is there a unity?

Task 2 What is the additive inverse of a? Why?

Task 3 Is $\oplus$ associative? Is $\otimes$ associative?

Task 4 Show that $\otimes$ is distributive over $\oplus$.

SUMMARY OF RINGS AND THEIR PROPERTIES

Ring	Form of Element	Unity	Abelian	Integral Domain	Field
$\mathbb{Z}$	k	1	Yes	Yes	No
$\mathbb{Z}_n$, n composite					
$\mathbb{Z}_p$, p prime					
$\mathbb{Z}[x]$					
$n\mathbb{Z}$, $n > 1$					
$M(\mathbb{Z})$, 2×2 matrices					
$\mathbb{Z}[i]$					
$\mathbb{Z}_3[i]$					
$\mathbb{Z}_2[i]$					
$\mathbb{Z}[\sqrt{2}]$					
$\mathbb{Q}[\sqrt{2}]$	$a + b\sqrt{2}$				
$\mathbb{Z} \oplus \mathbb{Z}$					

$\mathbb{Z}_3[i]$ IS A FIELD WITH 9 ELEMENTS

This field consists of all elements of the form $m + ni$ where m and n are in {0,1,2}. The multiplication table is:

	1	2	i	$1+i$	$2+i$	$2i$	$1+2i$	$2+2i$
1	1	2	i	$1+i$	$2+i$	$2i$	$1+2i$	$2+2i$
2	2	1	$2i$	$2+2i$	$1+2i$	i	$2+i$	$1+i$
i	i	$2i$						
$1+i$	$1+i$	$2+2i$						
$2+i$	$2+i$	$1+2i$						
$2i$	$2i$	i						
$1+2i$	$1+2i$	$2+i$						
$2+2i$	$2+2i$	$1+i$						

Task 1 Complete the table of inverses and orders:

x	1	2	i	$1+i$	$2+i$	$2i$	$1+2i$	$2+2i$
x^{-1}								
order of x								

Task 2 Which familiar group is the multiplicative group isomorphic to?

APPLICATION OF A FAMOUS THEOREM

PROBLEM: Explain why $x^7 - x = x(x-1)(x-2)(x-3)(x-4)(x-5)(x-6)$ in $\mathbb{Z}_7[x]$.

There are a number of ways of approaching this problem, several motivated by a similar question you couild ask in the 8th grade. Explain why $x^2 - 5x + 6 = (x-2)(x-3)$. The following tasks guide you through three solutions.

Task 1 Use a calculator and show that each of 1, 2, 3, 4, 5, 6 satisfies $x^7 - x = 0$; then use the factor theorem.

Task 2 Simplify the right hand side algebraically using $x - 6 = x + 1$, $x - 5 = x + 2$, $x - 4 = x + 3$.

Task 3 What famous Theorem has $x^7 - x \equiv 0$ or $x^7 \equiv x$ in it?

Made in the USA
Las Vegas, NV
26 August 2021

28975554R00103